Springer
Berlin
Heidelberg
New York
Barcelona
Hong Kong
London
Milan
Paris
Singapore
Tokyo

Applications of Mathematics

Guy Fayolle
Roudolf Iasnogorodski
Vadim Malyshev

Random Walks
in the Quarter-Plane

Algebraic Methods,
Boundary Value Problems and Applications

Springer

Guy Fayolle

INRIA, Domaine de Voluceau, Rocquencourt
F-78153 Le Chesnay, France
e-mail: guy.fayolle@inria.fr

Roudolf Iasnogorodski

University of Orléans
Department of Mathematics
F-45046 Orléans la Source, France
e-mail: roudolf.iasnogorodski@labomath.univ-orleans.fr

Vadim Malyshev

INRIA, Domaine de Voluceau, Rocquencourt
F-78153 Le Chesnay, France
e-mail: vadim.malyshev@inria.fr

Managing Editors

I. Karatzas
Departments of Mathematics
and Statistics
Columbia University
New York, NY 10027, USA

M. Yor
CNRS, Laboratoire de Probabilités
Université Pierre et Marie Curie
4 Place Jussieu, Tour 56
F-75230 Paris Cedex 05, France

Mathematics Subject Classification (1991):
60J15, 30D05, 35Q15, 30F10, 60K25

Library of Congress Cataloging-in-Publication Data

Fayolle, G. (Guy), 1943-
    Random walks in the quarter-plane : algebraic methods, boundary
value problems, and applications / Guy Fayolle, Roudolf
Iasnogorodski, Vadim Malyshev.
        p.   cm. -- (Applications of mathematics, ISSN 0172-4568 ; 40)
    Includes bibliographical references and index.
    ISBN-13:978-642-64217-3   e-ISBN-13:978-3-642-60001-2
    DOI: 10.1007/978-3-642-60001-2

    1. Random walks (Mathematics)   I. Iasnogorodski, Roudolf, 1938-
. II. Malyshev, V. A. (Vadim Aleksandrovich)   III. Title.
IV. Series.
QA274.73.F39   1999
519.2'82--dc21   98-52367
                                                               CIP

ISSN 0172-4568
ISBN-13:978-642-64217-3   Springer-Verlag Berlin Heidelberg New York

© Springer-Verlag Berlin Heidelberg 1999
Softcover reprint of the hardcover 1st edition 1999

Typeset from the authors' LaTeX files using Springer-TeX style files
SPIN: 10078403   41/3143 - 5 4 3 2 1 0 – Printed on acid-free paper

Guy Fayolle: *To my father, who has just lost his fight for life*

Roudolf Iasnogorodski: *To my mother, who would have liked to see this book*

Vadim Malyshev: *To my parents, who taught me to be open with people, even at my own expense*

Introduction

Historical Comments

Two-dimensional random walks in domains with non-smooth boundaries interest several groups of the mathematical community. In fact these objects are encountered in pure probabilistic problems, as well as in applications involving queueing theory. This monograph aims at promoting original mathematical methods to determine the invariant measure of such processes. Moreover, as it will emerge later, these methods can also be employed to characterize the transient behavior. It is worth to place our work in its historical context.

This book has three sources.

1. Boundary value problems for functions of one complex variable;

2. Singular integral equations, Wiener-Hopf equations, Toeplitz operators;

3. Random walks on a half-line and related queueing problems.

The first two topics were for a long time in the center of interest of many well known mathematicians: Riemann, Sokhotski, Hilbert, Plemelj, Carleman, Wiener, Hopf. This *one-dimensional theory* took its final form in the works of Krein, Muskhelishvili, Gakhov, Gokhberg, etc.

The third point, and the related probabilistic problems, have been thoroughly investigated by Spitzer, Feller, Baxter, Borovkov, Cohen, etc.

To sketch the march of thought, let us recall that many simple problems pertaining to the $M/M/k$ queue amount, very roughly, to solve equations of the form

$$Q(z)\pi(z) = U(z), \quad \forall |z| \leq 1,$$

where U and π are unknown, holomorphic in the unit disk $\mathcal{D}$, continuous on the unit circle; U is a polynomial of degree m, and π is the generating function of a discrete positive random variable, say the number of units in the system. Then, provided the model is meaningful, the known function Q has in general m zeros in $\mathcal{D}$ and this sufficient to determine U and π.

In the $GI/GI/1$ queue, a basic problem was to find the distribution of the stationary waiting time W of an arriving customer, which satisfies the stochastic equation

$$W \stackrel{\mathcal{L}aw}{=} [W + \sigma]^+, \qquad (0.0.1)$$

where σ is a known random variable, independent of W and taking real values. Using Laplace transforms, equation (0.0.1) yields

$$\omega^-(s) = \gamma(s)\omega^+(s), \qquad (0.0.2)$$

where $\gamma(s) = 1 - E[e^{-s\sigma}]$ and the unknown functions ω^+, ω^- are holomorphic in the right [resp. left] half-plane. The function γ is given, but a priori defined only on the imaginary axis. It turns out that (0.0.2) is exactly equivalent to the famous *Wiener-Hopf factorization* problem.

After this, there were two main directions of generalization for multidimensional situations.

• The first one is a multidimensional analog of the Wiener-Hopf factorization, when the problem or equation depends on extra-parameters. The theory in then is quite analogous to the one-dimensional case and the simplest example concerns Wiener-Hopf equations in a half-space. Much more involved is the index theory of differential and pseudo-differential operators on manifolds with smooth boundaries. However, when one wants to get index theory for Wiener-Hopf equations in an orthant of dimension k or other domain with piecewise smooth boundaries, it becomes immediately clear that one needs more knowledge for similar equations in dimension $k - 1$. For example, there is an index theory for equations in a quarter plane ($k = 2$), due to B. Simonenko, because we have a complete control for $k = 1$.

• Then, completely new approaches to these problems were discovered by the authors of this book, the goal going much beyond the mere obtention of an index theory for the quarter plane. The main results can be summarized as follows.

1. Use of generating functions, or Laplace transforms, is quite standard and the resulting functional equations give rise to boundary value problems involving two complex variables. When the jumps of the random walk are bounded by 1 in the interior of the quarter plane, the basic functional equation writes

$$Q(x, y)\pi(x, y) + q(x, y)\pi(x) + \widetilde{q}(x, y)\widetilde{\pi}(y) + q_0(x, y)\pi_{00} = 0. \qquad (0.0.3)$$

2. The first step, quite similar to a Wiener-Hopf factorization, consists in considering the above equation on the algebraic curve $Q = 0$ (which is *elliptic* in the generic situation), so that we are left then with an equation for two unknown functions of one variable on this curve.

3. Next a crucial idea is to use Galois automorphisms on this algebraic curve in order to solve equation (0.0.3). It is clear that we need more information, which is obtained by using the fact that the unknown functions π and $\widetilde{\pi}$ depend respectively solely on x and y, i.e. they are invariant with respect to the corresponding Galois automorphisms of the algebraic curve. It is then possible to prove that the unknown π and $\widetilde{\pi}$ can be *lifted* as meromorphic functions onto the *universal covering* of some Riemann surface **S**. Here **S** corresponds to the algebraic curve $Q = 0$ and the universal covering is the complex plane $\mathbb{C}$.

4. Lifted onto the universal covering, π (and also $\widetilde{\pi}$) satifies a system of non-local equations having the simple form

$$\begin{cases} \pi(t + \omega_1) = \pi(t), & \forall t \in \mathbb{C}, \\ \pi(t + \omega_3) = a(t)\pi(t) + b(t), & \forall t \in \mathbb{C}, \end{cases}$$

where ω_1 [resp. ω_3] is a complex [resp. real] constant. The solution can be presented in terms of infinite series equivalently to *Abelian integrals*. The backward transformation (projection) from the universal covering onto the initial coordinates can be given in terms of uniformization functions, which, in the circumstances, are elliptic functions.

5. Another direct approach to solve the fundamental equation consists in working solely in the complex plane. After making the analytic continuation, one shows that the determination of π reduces to a boundary value problem (BVP), belonging to the Riemann-Hilbert-Carleman class, the basic form of which can be formulated as follows.

Let $\mathcal{L}^+$ denote the interior of domain bounded by a simple smooth closed contour $\mathcal{L}$.

Find a function Φ^+ holomorphic in $\mathcal{L}^+$, the limiting values of which are continuous on the contour and satisfy the relation

$$\Phi^+(\alpha(t)) = G(t)\Phi^+(t) + g(t), \quad t \in \mathcal{L},$$

where

* $g, G \in \mathbb{H}_\mu(\mathcal{L})$ (*Hölder condition* with parameter μ on $\mathcal{L}$);
* α, referred to as a *shift* in the sequel, is a function establishing a one to one mapping of the contour $\mathcal{L}$ onto itself, such that the direction of traversing $\mathcal{L}$ is changed and

$$\alpha'(t) = \frac{d\alpha(t)}{dt} \in \mathbb{H}_\mu(\mathcal{L}), \quad \alpha'(t) \neq 0, \quad \forall t \in \mathcal{L}.$$

In addition, most of the time, the function α is subject to the so-called *Carleman's condition*

$$\alpha(\alpha(t)) = t, \quad \forall t \in \mathcal{L}, \quad \text{where typically} \quad \alpha(t) = \bar{t}.$$

6. Analytic continuation gives clear understanding of possible singularities and thus allows to get the asymptotics of the solution

All these techniques work quite similarly for Toeplitz operators or for random walk problems. Here our presentation is given for random walks only for the sake of concreteness. More exactly, we consider the problem of calculating stationary probabilities for ergodic random walks in a quarter plane. But other problems for such random walks can be treated as well: transient behavior, first hitting problem [36] and calculating the Martin boundary.

The approach relating to points 1, 2, 3, 4 was mainly settled in the period 1968-1972 (see e.g. [46], [48], [49], [47], [57]).

The method in point 5 was proposed in the fundamental study [26], carried out in 1977-1979, which was strongly referred and pursued in many other papers (see [6], [7], [8], [10], [12], [13], [17], [20], [28], [23], [27], [24], [25], [30], [60], [63], [64]) and in the book by Cohen and Boxma [19].

Contents of the Book

In chapter 1, it is explained how the functional equations appear and why they bring complete knowledge about the initial problem. Section 2.1 contains some material necessary for the rest of the book. In section 2.2, the first step is presented, namely the restriction of the equation to the algebraic curve. This curve is studied in section 2.3 and 2.5, together with the initial domains of analyticity of the unknown functions. Simple basic properties of our Galois automorphisms are given in section 2.4,where the notion of the *group* of the random walk is introduced.

Chapter 3 is exclusively devoted to the analytic continuation of the unknown functions.

When the group of the random walk is finite, a very beautiful algebraic theory exists, for solving the fundamental equations (both the homogeneous and non-homogeneous one). This is the subject of chapter 4. We get necessary and sufficient conditions for the solution to be rational or algebraic. In fact, the solution is obtained in the general case. The analysis contains completely new results belonging to the authors, which are first published here. It is also interesting to note that finiteness of the group takes place for some simple queueing systems, which are discussed here as well. A first concrete example of algebraic solution in a queueing context was found in [31].

The case of an arbitrary group is made in chapter 5, by reduction to a Riemann-Hilbert BVP in the complex plane. Necessary and sufficient conditions for the process to be ergodic are obtained in a purely analytical way, by calculating the index of the BVP (which, roughly speaking, gives the number of independent admissible solutions). The unknown functions are given by integral forms, which can be explicitly computed, via the Weierstrass $\wp$-function.

Chapter 6 concerns degenerate but practically important cases (e.g. priority queues, joining the shorter of two queues, etc.), when the genus of the algebraic curve is zero. [32] In this case even simpler solutions exist.

Clearly, we could not cover in this book all related problems and ideas. Some of them, concerning in particular the asymptotic behavior of the probability distribution, are discussed in the final chapter 7.

Acknowledgements

The authors are indebted to Martine Verneuille, who kindly managed to transform into LaTeX a great deal of a manuscript, which was sometimes - and this is an understatement - not easy to decipher, and to Angela Thomin at Springer-Verlag for her assistance.

Table of Contents

1. Probabilistic Background

1.1 Markov Chains

For this brief section, the readers who are not probabilists are referred to any standard text-book on countable Markov chains (MC) in discrete time. We shall give exactly the altogether indispensable amount of information which is needed.

A discrete time homogeneous Markov chain with a denumerable state space $\mathcal{A}$ is defined by the stochastic matrix

$$\mathbf{P} = \|p_{\alpha\beta}\|, \quad \alpha, \beta \in \mathcal{A},$$

such that

$$p_{\alpha\beta} \geq 0, \quad \sum_{\beta} p_{\alpha\beta} = 1, \quad \forall \alpha \in \mathcal{A}.$$

The matrix elements of $\mathbf{P}^n$ will be denoted by $p_{\alpha\beta}^{(n)}$.

Definition 1.1.1 *A MC is called* irreducible *if, for any ordered pair α, β, there exits m, depending on (α, β), such that*

$$p_{\alpha\beta}^{(m)} \neq 0.$$

In addition, an irreducible MC is called *aperiodic* if, for some $\alpha, \beta \in \mathcal{A}$, the set $\{n : p_{\alpha\beta}^{(n)} \neq 0\}$ has a greatest common divisor equal to 1. It follows that the same property is true for all ordered pairs α, β.

Definition 1.1.2 *An irreducible aperiodic MC is called* ergodic *if, and only if, the equation*

$$\pi\mathbf{P} = \pi, \tag{1.1.1}$$

where π is the row vector $\pi = (\pi_\alpha, \alpha \in \mathcal{A})$, has a unique ℓ_1-solution up to a multiplicative factor, which can be chosen so that

$$\sum_{\alpha} \pi_\alpha = 1, \quad \text{together with } \pi_\alpha > 0.$$

The π_α's are called stationary probabilities. See [11]. *The random variable representing the position of the chain at time n will be written X_n and X will denote the random variable with distribution π.*

1.2 Random Walks in a Quarter Plane

The class of MC we shall mainly consider in this book are called *maximally space homogeneous random walks*. They are characterized by the three following properties.

P1 The state space is $\mathcal{A} = \mathbb{Z}_+^2 = \{(i,j) : i,j \geq 0 \text{ are integers}\}$.

P2 *(Maximal space homogeneity)* $\mathbb{Z}_+^2$ is supposed to be represented as the union of a finite number of non-intersecting classes

$$\mathbb{Z}_+^2 = \bigcup_r S_r. \tag{1.2.1}$$

Moreover, for each r and for all $\alpha \in S_r$ such that

$$p_{\alpha,\alpha+(i,j)} \neq 0, \quad \alpha + (i,j) \in \mathbb{Z}_+^2,$$

$p_{\alpha,\alpha+(i,j)}$ does not depend on α, and can therefore be denoted by $_r p_{ij}$. Throughout most of the book, the classes S_r will have the following structure:

$$\mathbb{Z}_+^2 = S \cup S' \cup S'' \cup \{(0,0)\} \tag{1.2.2}$$

where

$$\begin{cases} S &= \{(i,j) : i,j > 0\}, \\ S' &= \{(i,0) : i > 0\}, \\ S'' &= \{(0,j) : j > 0\}. \end{cases}$$

The *internal* parts S' and S'' are called respectively the x-axis and y-axis. In this case, the probabilities $_r p_{i,j}$ will be simply written p_{ij}, p'_{ij}, p''_{ij}, p^0_{ij}, according to their respective regions S, S', S'', and $\{(0,0)\}$. It is worth noting, nevertheless, that in section 1.3 a more general partition of the state space is considered.

The last property deals with the boundedness of the jumps, which will be supposed unless otherwise stated (see e.g. chapters 5 and 6).

P3 *(Boundedness of the jumps)* For any $\alpha \in S_r$,

$$p_{\alpha\beta} = 0, \text{ unless } -d_r^- \leq (\beta - \alpha)_i \leq d_r^+,$$

for some constants $0 \leq d_r^\pm < \infty$, where $(\beta - \alpha)_i$ is the i-th coordinate of the vector $\beta - \alpha$, $i = 1,2$. In addition, the next important assumption will hold throughout this book

$$d_r^\pm = 1, \quad \text{for the class } S_r = S.$$

The ergodicity conditions for the r.w. $\mathcal{L}$ can be given in terms of the mean jump vectors

$$\begin{cases} \mathbf{M} & = (M_x, M_y) = \left(\sum i p_{ij}, \sum j p_{ij} \right) \\ \mathbf{M'} & = (M_x', M_y') = \left(\sum i p_{ij}', \sum j p_{ij}' \right) \\ \mathbf{M''} & = (M_x'', M_y'') = \left(\sum i p_{ij}'', \sum j p_j'' \right). \end{cases} \tag{1.2.3}$$

We shall consider only *irreducible aperiodic random walks.*

Theorem 1.2.1   *When* $\mathbf{M} \neq 0$, *the random walk is ergodic if, and only if, one of the following three conditions holds:*

(i)
$$\begin{cases} M_x < 0, \; M_y < 0, \\ M_x M_y' - M_y M_x' < 0, \\ M_y M_x'' - M_x M_y'' < 0; \end{cases}$$

(ii) $M_x < 0, \quad M_y \geq 0, \quad M_y M_x'' - M_x M_y'' < 0;$

(iii) $M_x \geq 0, \quad M_y < 0, \quad M_x M_y' - M_y M_x' < 0.$

■

A probabilistic proof of this theorem exists in [29]. A new and purely analytic proof is presented in chapter 5 and the analysis of the case $\mathbf{M} = 0$ is carried out in detail in chapter 6.

As announced in the general introduction, this monograph intends to provide a methodology of an essentially analytic nature, for constructing and effectively computing the invariant measures associated to the random walks introduced in the present section. In fact, it is worth to emphasize that all these methods can be also employed (up to some additional technicalities) to analyze the transient behavior of the random walk, and to solve explicitly the classically so-called *Kolmogorov's equations*, which describe the time-evolution of the semigroup associated to a Markov process, see e.g. [11].

1.3 Functional Equations for the Invariant Measure

We derive here the fundamental functional equations to be used throughout the book. It is useful to present them in a more general situation, which means that for a while we do not assume any boundedness of the jumps. To that end, consider the MC

$$X_n = \left(X_n^1, \ldots, X_n^k \right), \quad n \geq 0, \quad X_n \in \mathbb{Z}_+^k,$$

with state space

$$\mathbb{Z}_+^k = \{ z = (z_1, \ldots, z_k) : z_i \geq 0, \quad i = 1, \ldots, k \}$$

which is partitioned into a finite number of classes

$$\mathbb{Z}_+^k = \bigcup S_r,$$

so that the following assumption holds: *two states belong to the same class S_r if, and only if, the probability distributions P_r of the jumps from these states are the same.* The corresponding probability densities are the $_rp_{ij}$ introduced in the preceding section 1.2.

Let us define the vector of complex variables

$$u = (u_1, \dots, u_k), \quad u_i \in \mathbb{C}, \quad |u_i| = 1, \quad i = 1, \dots, k,$$

and the jump generating functions

$$P_r(u) = E[u^{(X_{n+1}-X_n)}/X_n = z], \quad z \in S_r, \tag{1.3.1}$$

with the standard notation

$$u^z = \prod_{i=1}^{k} u_i^{z_i}.$$

Since by our assumptions $P_r(u)$ does not depend on z, Kolmogorov's equations take the form

$$E[u^{X_{n+1}}] = E\left[u^{X_n}u^{X_{n+1}-X_n}\right] = \sum_r E\left[u^{X_n}\mathbb{1}_{\{X_n \in S_r\}}\right] P_r(u). \tag{1.3.2}$$

To account for the stationary case, we introduce the generating functions

$$\pi_r(u) = E\left[u^X \mathbb{1}_{\{X \in S_r\}}\right] = \sum_{z \in S_r} \pi_z u^z, \tag{1.3.3}$$

where π_z denotes the stationary probability of being in state z. Taking the limit $n \to \infty$ in (1.3.2) and using (1.3.3), we get the basic equation

$$\sum_r [1 - P_r(u)] \pi_r(u) = 0. \tag{1.3.4}$$

Note that when the jumps are bounded from below, (1.3.4) is defined for all $u = (u_1, \dots, u_k)$, $u_i \in \mathbb{C}$, $|u_i| \leq 1$, $i = 1, \dots, k$. Since we shall mainly consider the case $k = 2$, it will be convenient to rewrite (1.3.4) in a more explicit way, by means of the notation below, which will be ubiquitous throughout the book.

$$\begin{cases}
\pi(x,y) & = \displaystyle\sum_{i,j=1}^{\infty} \pi_{ij} x^{i-1} y^{j-1}, \\[2mm]
\pi(x) & = \displaystyle\sum_{i\geq 1} \pi_{i0} x^{i-1}, \\[2mm]
\widetilde{\pi}(y) & = \displaystyle\sum_{j\geq 1} \pi_{0j} y^{j-1}, \\[2mm]
Q(x,y) & = xy\Big(-1+\displaystyle\sum_{i,j} p_{ij} x^i y^j\Big), \\[2mm]
q(x,y) & = x\Big(\displaystyle\sum_{i\geq -1}\sum_{j\geq 0} p'_{ij} x^i y^j - 1\Big) \\[2mm]
\widetilde{q}(x,y) & = y\Big(\displaystyle\sum_{i\geq 0}\sum_{j\geq -1} p''_{ij} x^i y^j - 1\Big), \\[2mm]
q_0(x,y) & = \Big(\displaystyle\sum_{i,j} p^0_{ij} x^i y^j - 1\Big),
\end{cases} \qquad (1.3.5)$$

where we have set $p^0_{ij} \overset{\text{def}}{=} P_{(0,0),(i,j)}$.

Now equation (1.3.4) takes the fundamental form

$$\boxed{-Q(x,y)\pi(x,y) = q(x,y)\pi(x) + \widetilde{q}(x,y)\widetilde{\pi}(y) + \pi_{00}q_0(x,y).} \qquad (1.3.6)$$

When property **P3** holds, it is immediate to check that the functions $Q, q, \widetilde{q}$ and q_0 introduced in (1.3.5) are polynomials in x, y. In addition, $\pi(x,y), \pi(x), \pi(y)$ have to be holomorphic in the region $|x|, |y| < 1$. Thus, the analysis of the invariant measure of the random walk amounts to solve the functional equation (1.3.6), in agreement with the next theorem.

Theorem 1.3.1   *For the irreducible aperiodic random walk to be ergodic, it is necessary and sufficient that there exist $\pi(x,y), \pi(x), \pi(y)$ holomorphic in $|x|, |y| < 1$, and a constant π_{00}, satisfying the fundamental equation (1.3.6) together with the ℓ_1-condition*

$$\sum_{i,j=0}^{\infty} |\pi_{ij}| < \infty. \qquad (1.3.7)$$

In this case these functions are unique.   ■

Theorem 1.3.1 proceeds directly from the material given in definition 1.1.2, asserting existence and uniqueness of a finite invariant measure for irreducible ergodic Markov chains. We shall seek for solutions of (1.3.6) from the following point of view:

Find functions $\pi(x,y)$, $\pi(x)$, $\widetilde{\pi}(y)$, satisfying (1.3.6), holomorphic in $\mathcal{D} \times \mathcal{D}$ and continuous in $\overline{\mathcal{D}} \times \overline{\mathcal{D}}$, where

$$\mathcal{D} \stackrel{def}{=} \{z \in \mathbb{C} : |z| < 1\} \quad \text{and} \quad \overline{\mathcal{D}} \stackrel{def}{=} \{z \in \mathbb{C} : |z| \leq 1\}.$$

The main idea consists in working on the variety $Q(x, y) = 0, (x, y) \in \overline{\mathcal{D}} \times \overline{\mathcal{D}}$ and the content of the book shows, by means of various approaches, that this is sufficient to obtain all aforementioned functions.

Remark 1.3.2 A priori, *finding a solution* $\pi(x, y)$, *holomorphic in* $\mathcal{D} \times \mathcal{D}$ *and continuous in* $\overline{\mathcal{D}} \times \overline{\mathcal{D}}$, *does not imply the* ℓ_1-*condition (1.3.7), as it emerges from the theory of functions of several complex variables (see for instance [9]). Furthermore, supposing the system is not ergodic, one will see that a solution of (1.3.6) can yet exist, holomorphic in* $\mathcal{D}_a \times \mathcal{D}_a$, *with* $a < 1$, *where* $\mathcal{D}_a$ *is the disk* $\mathcal{D}_a \stackrel{def}{=} \{z \in \mathbb{C} : |z| < a\}$.

2. Foundations of the Analytic Approach

2.1 Fundamental Notions and Definitions

We gather here some of the basic notions which appear elsewhere in the book and are strictly necessary for our purpose, namely the resolution of the equations (1.3.6). The logical structure of this paragraph is self contained, but, for a deeper and more detailed presentation, we refer the reader to classical monographs [34, 72]. Although the first five chapters of the book demand much less generality, we think that a more abstract understanding is nevertheless useful.

A separable topological space is called a *two-dimensional manifold* $\mathbb{M}$, if each point belongs to a neighborhood which is homeomorphic to an open disk in the complex plane $\mathbb{C}$. A pair (U, φ), formed by some neighborhood $U \subset \mathbb{M}$ and its associated homeomorphism φ is called a *chart*. The mapping $\varphi : U \to \mathbb{C}$ defines a system of local coordinates in U. A collection of charts $\{(U_i, \varphi_i), \ i \in I\}$, where, for some index set I, $\{U_i, \ i \in I\}$ is an open covering of $\mathbb{M}$, is called an *atlas* $\mathcal{A}$.

A connected two-dimensional manifold $\mathbb{M}$ is an (abstract) *Riemann surface* S, if there exists an atlas $\mathcal{A}_S$ with the following property:

> *For any pair (U, φ), (V, ψ) of charts in $\mathcal{A}_S$, such that $U \cap V \neq \emptyset$, the mapping $\varphi \circ \psi^{-1}$ is holomorphic in $\psi(U \cap V) \subset \mathbb{C}$.*

The classical notion of holomorphic functions can be generalized to the case of Riemann surfaces. Let S be a Riemann surface, $\mathcal{A}_S$ its atlas, and $Y \subset S$ an open connected set of S. A function $f : Y \to \mathbb{C}$ is said to be *holomorphic in Y*, if, for any chart (U, φ) in $\mathcal{A}_S$, the mapping $f \circ \varphi^{-1} : \varphi(U) \to \mathbb{C}$ is holomorphic in the normal sense in the open set $\varphi(U) \subset \mathbb{C}$. The set of all functions holomorphic in Y builds an algebra over $\mathbb{C}$.

Let now S and T be two Riemann surfaces. The mapping $f : S \to T$ is said to be holomorphic if, for any pair of charts (U, φ), (V, ψ) belonging to $\mathcal{A}_S$ and $\mathcal{A}_T$ respectively, with $f(U) \subset V$, the mapping $\psi \circ f \circ \varphi^{-1}$ is holomorphic in $\varphi(U) \subset \mathbb{C}$. The uniqueness theorem remains valid: if f_1 and f_2 are two holomorphic functions from S to T which are equal on an infinite compact set of S, then they must be equal everywhere.

In a similar way, the definition for a function to be meromorphic can be extended as follows. Let $Y \subset S$ an open set of the Riemann surface S. A function f :

$Y \to \mathbb{C}$ is called *meromorphic* in Y if f is holomorphic in $Y - Y_0$, where $Y_0 \subset Y$ is a set of isolated points such that

$$\lim_{x \in Y, x \to p} |f(x)| = \infty, \quad \forall p \in Y_0.$$

The points of Y_0 are called poles of f. Note that one can show that locally f is the quotient of two holomorphic functions. The set of meromorphic functions in Y form a field. It is classical to remark that, after *compactification*, the complex plane $\mathbb{C}$ is isomorphic to the Riemann sphere $\mathbb{P}^1$, the point at infinity becomes an ordinary point and the meromorphic functions on Y can be considered as holomorphic mappings from Y to $\mathbb{P}^1$.

2.1.1 Covering Manifolds

Let X and Y represent two Riemann surfaces and $h : Y \to X$ a holomorphic mapping of Y *onto* X (i.e. to any point of X, there corresponds at least one point of Y). This assumption is useful only when Y is not compact, since it can be shown that, whenever Y is compact, any holomorphic mapping h is necessarily *onto* and that X is compact [34]. The pair (Y, h) is called a *cover* and Y is a *covering manifold* of X. When h is not constant, $h^{-1}(x)$ is a discrete subset of Y, for any $x \in X$. A point $y \in h^{-1}(x)$ is said to *lie over x* (or to *cover x*) and x is the *projection of y* on X.

Definition 2.1.1 *If, for any neighborhood V of y, the restricted mapping $h_{|V}$ is not one-to-one, y is called a* branch point. *More precisely, $y \in V$ is a* branch point of order $n - 1$ *if there exist two charts $(U, \varphi) \in A_X$, $(V, \psi) \in A_Y$, $y \in V$, with $h(V) \subset U$, and a finite number n, such that*

$$(\varphi \circ h)(v) \equiv \gamma_n \circ \psi(v), \quad \forall v \in V,$$

where the function $\gamma_n : \mathbb{C} \to \mathbb{C}$ satisfies $\gamma_n(z) = z^n$. This means that each point $x \in h(V) \setminus h(y)$ is covered by exactly n points of Y contained in $V \setminus y$, the deleted neighborhood of y. Often n will be referred to as the multiplicity. *If $n = 1$, i.e. h is a one-to-one mapping $V \to U$, for some V, then y is called a* regular *point.*

Since h is holomorphic, it can be shown that there exists an integer s, such that every point $x \in X$ is covered by exactly s points of Y, provided that a branch point of order $k - 1$ is counted k times. Then, the covering manifold Y will be called an *s-sheeted* covering of X. It is important to note that s can be infinite, and we shall encounter this case. Among all covering maps without branch points, there exists exactly one, up to an isomorphism, which is simply connected and usually referred to as *the universal covering*.

Definition 2.1.2 [Proper mapping] *Let (Y, h) be a cover of X. The mapping h is called* proper *if the preimage of every compact set is compact. In particular, if Y is compact then h is always proper.*

Suppose that X and Y are Riemann surfaces and $h : Y \to X$ is a proper non-constant holomorphic mapping. Let B be the set of branch-points of h and $A \stackrel{\text{def}}{=} h(B)$. Let also $Y' \stackrel{\text{def}}{=} Y \setminus h^{-1}(A)$ and $X' \stackrel{\text{def}}{=} X \setminus A$. Then the restriction $h_{|Y'} : Y' \to X'$ is a proper unbranched covering, which has a well defined finite number of sheets n. The following more general property is valid.

Proposition 2.1.3 *Let X and Y be Riemann surfaces and $h : Y \to X$ is a proper non-constant holomorphic mapping. Then there exists a natural number n such that h takes every value $c \in X$, counting multiplicities, n times.*

Definition 2.1.4 [Lifting] *Let us introduce the following objects.*

* X, Y, *two Riemann surfaces, and Z a topological space;*

* $h : Y \to X$, *a holomorphic mapping;*

* $f : Z \to X$, *a continuous function;*

A lifting of f with respect to h is a continuous mapping $g : Z \to Y$, such that $f = h \circ g$.

Proposition 2.1.5 *Let (Y, h) be a cover of X. Let L be an arc of X having its origin at x (i.e. there exists a continuous mapping $f : [0, 1] \to X$, where $L = f[0, 1]$ and $f(0) = x$). Then, $\forall y \in h^{-1}(x)$, $\exists$ an arc in Y, which has its origin at y and lies over L.*

Proposition 2.1.6 [Uniqueness of lifting]

(i) *Let (Y, h) be a cover without branch-point. Let g_1 and g_2 be two liftings with respect to h. Then either $g_1(z) \neq g_2(z)$, $\forall z \in Z$, or $g_1 \equiv g_2$. In addition, if f is holomorphic, then any lifting g is also holomorphic.*

(ii) *In general, when there are branch points, the preceding property holds provided that, with the notation introduced just after definition 2.1.2, X, Y and Z are replaced respectively by X', Y' and $Z' = f^{-1}(X')$.*

Remark 2.1.7 *This proposition will be used either to lift arcs, in which case $Z = [0, 1]$, or also to compare different mappings, taking then $Z \equiv Y$.*

The study of analytic functions in the complex plane entails frequently the introduction of *Riemann surfaces* on which the analytic functions are single-valued. In most of the cases –and particularly in this book– these surfaces are composed of several sheets lying over $\mathbb{C}$ and they are compatible with the definition of abstract Riemann surfaces given above. In fact, they can be described by the *gluing of germs*. The *germ* of a function f at a point $x \in \mathbb{C}$ is the set of all disks $\mathcal{O}_x$ with center x, such that there exists a convergent power series

$$f(z) = \sum_{n=0}^{\infty} a_n (z - x)^n, \quad \forall z \in \mathcal{O}_x. \tag{2.1.1}$$

In general, for given x and $\mathcal{O}_x$, there can exist k_x different series (2.1.1), each of them representing , for instance, the branches of an algebraic function (see definition thereafter). All the pairs $(\mathcal{O}_x, k_x)$, numbered in an arbitrary way, can be pasted together to build an *analytic configuration* which is the Riemann surface of the analytic function f.

The problem of uniqueness of the above analytic continuation process is solved by the following fundamental *monodromy theorem*.

Theorem 2.1.8 *Let X be a Riemann surface. Let u_0 and u_1 two homotopic curves on X, each joining the points A and B, which implies the existence of u_t, $t \in [0, 1]$, a set of curves representing a continuous deformation of u_0 into u_1. Let f be a function holomorphic in a neighborhood of A and admitting an analytic continuation over u_t, $\forall t \in [0, 1]$. Then the two analytic continuations, over u_0 and u_1 respectively, lead to the same holomorphic function in a neighborhood of B. In particular, if X is simply connected, then all closed curves on X are homotopic to one point and, if f can be continued along any curve starting from A, then there exists one and only one function holomorphic on X and coinciding with f in a neighborhood at A.* ∎

2.1.2 Algebraic Functions

Let X be a Riemann surface and $\mathcal{M}(X)$ be the field of meromorphic functions on X, and let

$$P(T) = \sum_{i=0}^{n} f_i T^i, \quad f_i \in \mathcal{M}(X),$$

be an irreducible polynomial of degree n in one indeterminate T, with coefficients in $\mathcal{M}(X)$. When X is the complex plane $\mathbb{C}$, the equation

$$P(Y) = \sum_{i=0}^{n} f_i Y^i = 0, \quad f_i \in \mathcal{M}(X). \tag{2.1.2}$$

implicitly defines a *multi-valued* function $Y(x)$, $x \in \mathbb{C}$, called an *algebraic function*. Equation (2.1.2) has in the neighborhood of every x, but a finite number of them called *critical points*, exactly n distinct finite roots $Y_1(x), Y_2(x), \ldots, Y_n(x)$, which are the *branches* of the algebraic function $Y(x)$. In the domain obtained by removing from the complex plane the critical points, each branch of the function $Y(x)$ is *regular*. There are any two sorts of critical points: *poles* and multiple points called *branch points*. This point at infinity in $\mathbb{C}$ does not play a special role and can be both a regular or a branch point, as can be seen by a change of variables.

At any regular point x_0, there exist exactly n different convergent power series (or germs)

$$Y_{x_0,k}(x) = \sum_{l=0}^{\infty} c_l(x_0, k)(x - x_0)^l, \quad k = 1, 2, \ldots n,$$

satisfying (2.1.2). Moreover, in the neighborhood of a branch point, the algebraic function is made of a *cyclical* system of branches, each of them having a power series expansion with fractional exponents. It is possible to construct a Riemann surface, referred to as the *Riemann surface of the algebraic function* $Y(x)$, by putting together the above germs. One of the main goal pursued by Riemann, when he introduced these abstract surfaces, was to render algebraic functions single-valued. This possibility is contained in the following theorem.

Theorem 2.1.9   *There exist an holomorphic n-sheeted proper cover* (Y, π) *of X and a function $F \in \mathcal{M}(Y)$, such that*

$$\sum_{i=0}^{n} f_i(\pi(y))(F(y))^i = 0, \quad \forall y \in Y. \tag{2.1.3}$$

The triple (Y, π, F) is unique up to an isomorphism and is sometimes called (see e.g [34]) the algebraic function defined by the polynomial P. ∎

Up to some notational abuse, Y is simply called the Riemann surface of P, and $Y(x)$ is the corresponding algebraic function. Frequently, X will be the Riemann sphere. In this situation Y is compact, since X is compact. It can be shown that every compact abstract Riemann surface can be realized as the Riemann surface of some algebraic function.

The above theorem is in fact intimately connected to Galois theory, the principles of which will be outlined now.

2.1.3 Elements of Galois Theory

Let K and L denote two arbitrary fields, such that $K \subset L$. We say that L is a *finite extension* of K, if L is a vector space over K of finite dimension n which is called the *degree of L over K* and is denoted by $[L : K]$.

Let $K[T]$ be the polynomial ring in one indeterminate T over the field K. The extension L of K is said to be *normal* or a *Galois extension* of K, if any polynomial $P \in K[T]$, irreducible and having a root in L, has all its roots in L. The necessary and sufficient condition for L to be normal is that there exists $P \in K[T]$ irreducible, such that the roots of P form a basis of L, as a vector space on K.

The *Galois group* $G(L/K)$ of the extension L of K is the group of automorphisms of L leaving all elements of K invariant. The order of this group is denoted by $[L:K]$ and any of its elements realizes a permutation of the roots of any polynomial $P \in K[T]$, provided that P is irreducible and has all its roots in L. Any extension of degree 2 is obviously normal.

Let (Y,π) be a cover of X. An automorphism σ of Y is called a *covering* automorphism if $\pi = \pi \circ \sigma$, which means that $\forall y \in Y$, the points y and $\sigma(y)$ have the same projection on X. The set of all covering automorphisms form a multiplicative group denoted by $\mathrm{Aut}(Y/X)$, which is a subgroup of the group of automorphisms of Y.

Definition 2.1.10 *Let (Y,π) be a cover of X without branch point. One says that π is normal or Galois if, for any $y_0, y_1 \in Y$, such that $\pi(y_0) = \pi(y_1)$, there exists $\sigma \in \mathrm{Aut}(Y/X)$, such that $\sigma(y_0) = y_1$. From proposition 2.1.6, it follows that σ is unique, setting $Z = Y$, $f = \pi$ and $\sigma = g$.*

Proposition 2.1.11 *Let (Y,π) be a universal cover of X (i.e. simply connected and without branch-point). Then π is Galois and $\mathrm{Aut}\,(Y/X)$ is isomorphic to the* fundamental group *of X (i.e. the set of homotopy classes of closed curves going through some arbitrary but fixed point). (See e.g. [34]).* ∎

Proposition 2.1.12 *Let (Y,π) a cover of X, the mapping π being proper. Then every covering transformation $\sigma' \in \mathrm{Aut}(Y'/X')$ can be extended to a covering transformation $\sigma \in \mathrm{Aut}(Y/X)$ and π will be called Galois if $(Y', \pi_{|Y'})$ is Galois, in the sense of definition 2.1.10.* ∎

Finally, the following important general theorem holds.

Theorem 2.1.13   *Let (Y,π,F) the algebraic function corresponding to the equation $P(T) = 0$, where P is an irreducible polynomial of degree n, having all its coefficients in $\mathcal{M}(X)$.*

Define the mapping $\pi^* : \mathcal{M}(X) \to \mathcal{M}(Y)$ by the relation

$$[\pi^*(f)]\,(y) = f(\pi(y)), \quad \forall y \in Y \text{ and } f \in \mathcal{M}(X).$$

(This means that the range of π^* is a subfield of $\mathcal{M}(Y)$ which can be, in this way, identified with $\mathcal{M}(X)$). Then the field $\mathcal{M}(Y)$ is an extension of degree n of $\mathcal{M}(X)$ and is isomorphic to the quotient field $\mathcal{M}(X)[T]/P(T)$. Moreover the groups $\mathrm{Aut}(Y/X)$ and $G(\mathcal{M}(Y)/\mathcal{M}(X))$ are isomorphic and the mapping

$$\phi : \sigma \to \phi_\sigma, \quad \text{with } \phi_\sigma(f) = f \circ \sigma^{-1}, \quad \forall f \in \mathcal{M}(Y).$$

is an automorphism of $\mathcal{M}(Y)$ leaving $\mathcal{M}(X)$ fixed. The covering π is Galois if, and only if, $\mathcal{M}(Y)$ is a Galois extension of $\mathcal{M}(X)$. ∎

2.1.4 Universal Cover and Uniformization

Every Riemann surface of a polynomial with coefficients in $\mathcal{M}(\mathbb{P}^1)$ is topologically isomorphic to a sphere with g *handles* attached to it. The number g is called the *genus* of the Riemann surface.

Let S be a compact Riemann surface with its associated polynomial Q and (Ω, λ) its universal covering. λ is unique up to an automorphism of Ω. It is known [38] that only three possible situations can take place.

(i) $\Omega = \mathbb{P}^1$, the Riemann sphere, then $g = 0$.

(ii) $\Omega = \mathbb{C}$, the (finite) complex plane, then $g = 1$.

(iii) $\Omega = \mathcal{D}$, then open unit disk, then $g > 1$.

The next result can be found for example in [72].

Proposition 2.1.14 *There exist two functions* $f, g \in \mathcal{M}(S)$ *such that*

(i) $Q(f, g) = 0$;

(ii) *any function belonging to* $\mathcal{M}(S)$ *is a rational function of* f *and* g.

∎

We shall also use the following, see e.g. [34].

Proposition 2.1.15 *Every meromorphic function on* $\mathbb{P}^1$ *is rational.* ∎

Remembering that $\lambda : \Omega \to S$, it follows that the functions $\widetilde{f}$ and $\widetilde{g}$, respectively given by
$$\widetilde{f} = f \circ \lambda, \quad \widetilde{g} = g \circ \lambda,$$
with f, g given in proposition 2.1.14, are meromorphic on Ω and invariant with respect to the group of covering automorphisms. It is customary to state the following.

Definition 2.1.16 *The pair of functions* $\widetilde{f}, \widetilde{g}$ *gives an uniformization of* $\mathcal{M}(S)$, *or, equivalently, uniformizes any* $h \in \mathcal{M}(S)$.

2.1.5 Abelian Differentials and Divisors

Here we quickly present two essential notions, which will be essentially used in chapter 4. In general, any holomorphic or meromorphic differential on a Riemann surface S will be called an *abelian differential*.

1. A holomorphic differential on S is called *abelian differential of the first kind*.

2. A meromorphic differential all of whose singularities are poles of order ≥ 2 is called *an abelian differential of the second kind*.

3. The abelian differentials of the *third kind* are taken to be all abelian differentials on S.

Theorem 2.1.17 *[72] The vector space of holomorphic differentials on a compact Riemann surface has dimension equal to its genus g.*   ∎

In particular, when the genus is 1, Abelian differentials of the first kind are unique up to a multiplicative constant. See also remark 3.3.5 in section 3.3.

An important question is to be able to specify the locations and the orders of poles and zeros of meromorphic functions defined on an arbitrary compact Riemann surface S. To this end, one designates several points $P_1, P_2, \ldots, P_k$ and the associate orders (integers) $m_1, m_2, \ldots, m_k$ of these points. Then the following symbol

$$P_1^{m_1} P_2^{m_2} \ldots P_k^{m_k}$$

will be used to denote the points $P_1, P_2, \ldots, P_k$, with the corresponding associate integers, and will be called a *divisor*. When P_i is a zero (or pole) with multiplicity $m_i, i = 1, \ldots, k$, of some meromorphic function, the corresponding divisor is referred to as a *principal divisor*. For a divisor to be principal, a necessary condition is $\sum m_i = 0$ (see [72]). In chapter 4, we shall also need Abel's theorem stated hereafter for the sake of completeness.

Theorem 2.1.18 *[72] A necessary and sufficient condition for a divisor a to be a principal divisor is that there exists a singular 1-chain γ, with boundary $\partial \gamma$, such that*

$$\partial \gamma = \mathfrak{a}$$

and

$$\int_\gamma d\omega = 0,$$

for each differential of the first kind on S.   ∎

2.2 Restricting the Equation to an Algebraic Curve

In the preceding section, we have introduced a function $\pi(x, y)$ of two complex variables and two functions of one complex variable $\pi(x)$ and $\tilde{\pi}(x)$ which we would like to find. Here we present the method to get rid of $\pi(x, y)$ allowing us to write a functional equation involving only these two functions of one complex variables. This method has four facets, which can be easily converted into each other.

2.2.1 First Insight (Algebraic Functions)

Consider the multi-valued algebraic function $Y(x)$ satisfying the polynomial equation

$$Q(x, Y(x)) = 0, \quad \forall x \in \mathbb{C}, \tag{2.2.1}$$

where Q is given in (1.3.5). A first naïve approach consists just in instantiating $y = Y(x)$ into the fundamental equation (1.3.6. This substitution is a *priori* valid only for the pairs $(x, Y_i(x))$ such that

$$|x| \leq 1, \quad |Y_i(x)| \leq 1,$$

where $Y_i(x), i = 0, 1$, are the two branches of $Y(x)$. For these values, we get from (1.3.6)

$$0 = q(x, Y_i(x))\pi(x) + \widetilde{q}(x, Y_i(x))\widetilde{\pi}(Y_i(x)) + \pi_{00}q_0(x, Y_i(x)). \tag{2.2.2}$$

This approach will be used in a self-contained way, allowing to get and to solve an integral equation of Fredholm type for $\pi(x)$, by using convenient conformal transformations associated to elliptic functions.

2.2.2 Second Insight (Algebraic Curve)

Let $\Gamma_a \stackrel{\text{def}}{=} \{x \in \mathbb{C} : |x| = a\}$ with, for the sake of shortness,

$$\Gamma \stackrel{\text{def}}{=} \Gamma_1, \quad \text{the unit circle.}$$

Let $\mathcal{B}$ be the algebraic curve in $\mathbb{C}^2$ defined by the fundamental equation

$$Q(x, y) = 0. \tag{2.2.3}$$

As observed at the end of chapter 1, if there exist functions $\pi(x, y)$, $\pi(x)$, $\widetilde{\pi}(y)$, analytic in $\mathcal{D}^2 \equiv \mathcal{D} \times \mathcal{D}$, continuous in $\overline{\mathcal{D}}^2$ and satisfying the fundamental equation, then we have necessarily, for all $(x, y) \in \mathcal{B} \cap \mathcal{D}^2$,

$$q(x, y)\pi(x) + \widetilde{q}(x, y)\widetilde{\pi}(y) + q_0(x, y)\pi_{00} = 0. \tag{2.2.4}$$

But solutions of (2.2.4) a priori would not imply the continuity of $\pi(x, y)$ in $\overline{\mathcal{D}}^2$. That this continuity holds follows indeed from the next two lemmas.

Lemma 2.2.1   *Assume that conditions of theorem 1.2.1 hold and the polynomial $Q(x, y)$ is irreducible. Then $\exists \epsilon > 0$, such that the functions π and $\widetilde{\pi}$ can be analytically continued up to the circle $\Gamma_{1+\epsilon}$ in their respective complex plane. Moreover, in $\mathcal{D}_{1+\epsilon}^2$ they satisfy equation (2.2.4) in $\mathcal{B} \cap \mathcal{D}_{1+\epsilon}^2$.* ∎

This lemma will be proved later in section 2.5. Let us note that the second assertion follows directly from the first one by the principle of analytic continuation.

It is important to mention right away the case of zero drift, i.e. $\mathbf{M} = 0$, considered later on in section 6.5. Then the algebraic curve has genus zero, but the conditions of lemma 2.2.1 *do not hold*, since one will show that it is not possible to continue π nor $\widetilde{\pi}$ up to some $\Gamma_{1+\varepsilon}, \forall \varepsilon$.

Lemma 2.2.2   *Under the conditions of lemma 2.2.1 the function*

$$\pi(x,y) = -\frac{q(x,y)\pi(x) + \widetilde{q}(x,y)\widetilde{\pi}(y) + q_0(x,y)\pi_{00}}{Q(x,y)} \qquad (2.2.5)$$

is analytic in $\mathcal{D}_{1+\varepsilon}^2$.   ∎

Proof.   This is just a direct corollary of the famous Weierstrass *"Nullstellensatz"*, which claims that, if p is an irreducible polynomial, h is holomorphic in an open domain $V \subset \mathbb{C}^m$ and $h = 0$ on $\{p = 0\} \cap V$, then $\dfrac{h}{p}$ is holomorphic in V.   ∎

Remark 2.2.3   The reader has already guessed that this projection onto the algebraic curve can be carried out in more general situations and in particular for cases in dimension greater than two.

Remark 2.2.4   The analyticity of the functions $\pi(x,y)$, $\pi(x)$, $\widetilde{\pi}(x)$ in $\mathcal{D}_{1+\varepsilon}^2$ in the ergodic cases of theorem 1.2.1 can also be derived from a Lyapounov function method, by purely probabilistic arguments (see [29]). It can be shown, in particular, that there exist constants $C > 0, 0 < \alpha < 1$, which can depend on the parameters p_{ij}, p'_{ij}, p''_{ij}, such that

$$\pi_{ij} < C\alpha^{i+j}.$$

2.2.3 Third Insight (Factorization)

First, we show that, using a one-dimensional factorization of $Q(x,y)$ with respect to any of the two variables x, y, we obtain indeed the projection onto the algebraic curve. For this purpose, we have to introduce the two branches $Y_0(x)$ and $Y_1(x)$ of the algebraic function $Y(x)$ on the unit circle Γ. Exact definitions will be given in the next section.

Choosing $x \neq 1, |x| = 1$, we can write

$$Q(x,y) = a(x)Q^+(x,y)Q^-(x,y),$$

where

$$a(x) = p_{11}x^2 + p_{01}x + p_{-1,1}, \quad Q^+(x,y) = y(y - Y_1(x)), \quad Q^-(x,y) = 1 - \frac{Y_0(x)}{y}.$$

This is exactly a Wiener-Hopf factorization of Q considered as a function of y, after setting $y = \exp it$ for all real t. Then the fundamental equation (1.3.6) can be rewrittene as

$$a(x)Q^+(x,y)\pi(x,y) = \frac{q(x,y)\pi(x) + \tilde{q}(x,y)\tilde{\pi}(y) + \pi_{00}q_0(x,y)}{Q^-(x,y)}. \tag{2.2.6}$$

Let P_z^+ and P_z^- denote the projection operators in the commutative Banach algebra $\mathcal{B}$ of functions on the unit circle, that is

$$\mathcal{B} = \left\{ \sum_{k=-\infty}^{\infty} a_k z^k : \sum_{k=-\infty}^{\infty} |a_k| < \infty \right\},$$

$$P_z^+ \left(\sum_{-\infty}^{\infty} a_k z^k \right) = \sum_{k=0}^{\infty} a_k z^k, \quad P_z^- \left(\sum_{-\infty}^{\infty} a_k z^k \right) = \sum_{k=-\infty}^{-1} a_k z^k, \quad P_z^+ + P_z^- = I_d,$$

where I_d denotes the identity operator. We shall use the formula

$$P_z^- \left(\frac{\omega(z)}{1 - \frac{a}{z}} \right) = \omega(a)\frac{a}{z - a}, \quad \forall \omega \in P_z^+(\mathcal{B}), \quad |a| < 1. \tag{2.2.7}$$

Applying now P_y^- (taken as an operator acting on functions of y) to both sides of (2.2.6), using (2.2.7) with $a = Y_0(x)$ and simplifying by $\dfrac{Y_0(x)}{y - Y_0(x)}$, we get exactly equation (2.2.2) for the branch $Y_0(x)$.

The point $x = 1$ always requires a more careful analysis, which can be carried out by exploring the non stationary Kolmogorov equations and using generating functions in space-time variables.

2.2.4 Fourth Insight (Riemann Surfaces)

Assuming that the polynomial Q is irreducible, we know from section 2.1.2 that the equations

$$Q(x, Y) = 0 \quad \text{and} \quad Q(X, y) = 0$$

define two Riemann surfaces (with the two corresponding algebraic functions $Y(x)$ and $X(y)$). In fact (see e.g. [72]), these two surfaces have the same genus, and hence are conformally equivalent. Thus we will consider a *single* Riemann surface, denoted by $\mathbf{S}$, which is the Riemann surface for both $Y(x)$ and $X(y)$, but has two different coverings

$$h_x : \mathbf{S} \to \mathbb{C}_x, \quad h_y : \mathbf{S} \to \mathbb{C}_y,$$

where $\mathbb{C}_z$ denotes the complex plane with respect to the z-coordinate.

Any function f on a domain $V \subset \mathbb{C}_x$ can be *lifted* onto $h_x^{-1}(V) \subset \mathbf{S}$, yielding thus a new function $\hat{f} \stackrel{\text{def}}{=} f \circ h_x$. Correspondingly, one has $\hat{g} \stackrel{\text{def}}{=} g \circ h_y$. Hence, taking for f and g the identity functions, we are entitled to put

$$x(s) \stackrel{\text{def}}{=} h_x(s), \quad y(s) \stackrel{\text{def}}{=} h_y(s), \quad s \in \mathbf{S}. \tag{2.2.8}$$

The general flowchart is given in figure 2.2.1.

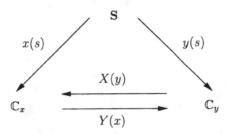

Fig. 2.2.1.

On $\mathbf{S}$, we have $Q(x(s), y(s)) = 0$ and it will be convenient to write

$$\hat{\pi}(s) \stackrel{\text{def}}{=} \pi(x(s)), \quad \hat{\tilde{\pi}}(s) \stackrel{\text{def}}{=} \tilde{\pi}(y(s)),$$
$$\hat{q}(s) \stackrel{\text{def}}{=} q(x(s), y(s)), \quad \text{etc.}$$

Then, for all $s \in h_x^{-1}(\mathcal{D}) \cap h_y^{-1}(\mathcal{D})$, we have the equation

$$\hat{\pi}(s)\hat{q}(s) + \hat{\tilde{\pi}}(s)\hat{\tilde{q}}(s) + \hat{q}_0(s) = 0, \tag{2.2.9}$$

in which $\hat{q}$, $\hat{\tilde{q}}$, $\hat{q}_0$ are meromorphic functions on $\mathbf{S}$, as resulting from the composition of meromorphic functions. We will reformulate lemma 2.2.1 on $\mathbf{S}$ in section 2.5. In fact, there are two constraints of primary importance: the solution $\hat{\pi}$ [resp. $\hat{\tilde{\pi}}$] of (2.2.9) has to be a function of the sole quantity $x(s)$ [resp. $y(s)$]. As shown in the next section, these properties of $\hat{\pi}$ and $\hat{\tilde{\pi}}$ will lead to the central idea, based on the use of *Galois* automorphisms of $\mathbf{S}$.

2.3 The Algebraic Curve $Q(x, y) = 0$

The present study relies essentially upon the properties of the algebraic functions $Y(x)$ and $X(y)$, defined by the curve in question. All necessary results will be put together and most of the features of these functions will be formulated in terms of the probabilities p_{ij} of the jumps in the interior of the quarter-plane.

Definition 2.3.1   *The random walk is called* singular *if the associated polynomial Q is either reducible or of degree 1 in at least one of the variables.*

Lemma 2.3.2   *The random walk is singular if, and only if, one of the following conditions holds:*

A *There exists $(i, j) \in \mathbb{Z}^2$, $|i| \leq 1$, $|j| \leq 1$, such that only p_{ij} and $p_{-i,-j}$ are different from 0 (see figure 2.3.1a and the three cases obtained by rotation);*

B *There exists i, $|i| = 1$, such that for any j, $|j| \leq 1$, $p_{ij} = 0$ (figure 2.3.1b,c);*

C *There exists j, $|j| = 1$, such that for any i, $|i| \leq 1$, $p_{ij} = 0$ (figure 2.3.1d,e).*

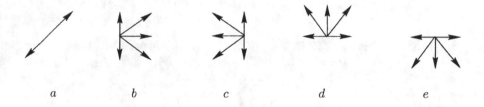

a b c d e

Fig. 2.3.1.

■

Proof.   Let us define the triple $a(x)$, $b(x)$, $c(x)$, [resp. $\tilde{a}(y), \tilde{b}(y), \tilde{c}(y)$] by

$$Q(x, y) \equiv xy\left(\sum p_{ij}x^i y^j - 1\right) = a(x)y^2 + b(x)y + c(x) = \tilde{a}(y)x^2 + \tilde{b}(y)x + \tilde{c}(y).$$

$$(2.3.1)$$

We assume $p_{00} < 1$, so that $b(x) \neq 0$, $\tilde{b}(y) \neq 0$. Then we have the following chain of equivalences:

$$Q \text{ is of degree 1 with respect to } y \Leftrightarrow a(x) = 0 \Leftrightarrow p_{i1} = 0, \quad \forall i.$$

This is exactly the cases **A** and **C** of the lemma, with $i = 1$ and $j = 1$ respectively. Analogously the two remaining cases in **B** and **C**, which correspond to $c(x) \equiv 0$ or $\tilde{c}(y) \equiv 0$, yield the reducibility of Q.

Suppose now that Q is of degree 2 both in x and in y. We shall consider two possibilities.

(a) $Q = f_1 f_2$, where f_1 and f_2 are of degree 1 in each variable. Then we can write, for $i = 1, 2$,

$$\begin{cases} f_i(x, y) & = & (A_i x + B_i)y - (C_i x + D_i), \\ (A_i, B_i) & \neq & (0, 0), \quad i = 1, 2, \\ (A_i, C_i) & \neq & (0, 0). \end{cases}$$

Ex hypothesis, $a(x) \not\equiv 0$. Since also $c(x) \not\equiv 0$ (as mentioned above), we have

$$\prod_{i=1}^{2}(A_i x + B_i) = a(x) \quad \text{and} \quad \prod_{i=1}^{2}(C_i x + D_i) = c(x),$$

which implies that the pair of vectors (A_1, B_1) and (A_2, B_2) [resp. (C_1, D_1), (C_2, D_2)] are simultaneously positive or negative. Thus, *ad libitum*, we can take choose (A_1, B_1) and (A_2, B_2) to be positive, i.e. $A_i, B_i \geq 0, i = 1, 2$.

The relationship

$$(A_1 x + B_1)(C_2 x + D_2) + (A_2 x + B_2)(C_1 x + D_1) = (1 - p_{00})x - p_{-1,0} - p_{10}x^2, \tag{2.3.2}$$

shows that the vectors (C_1, D_1) and (C_2, D_2) are both positive. This implies that $p_{-1,0} = p_{10} = 0$ and also that both polynomials $(A_1 x + B_1)(C_2 x + D_2)$ and $(A_2 x + B_2)(C_1 x + D_1)$ are linear in x. So we obtain

$$A_1 C_2 = B_1 D_2 = A_2 C_1 = B_2 D_1 = 0,$$

which yields either

$$B_1 = B_2 = C_1 = C_2 = 0$$

or

$$A_1 = A_2 = D_1 = D_2 = 0.$$

Consequently, there exists a polynomial p of one variable, such that $Q(x, y)$ is equal either to $p(xy)$ or to $x^2 p\left(\dfrac{y}{x}\right)$. Since $Q(1,1) = 0$, necessarily $p(1) = 0$. Here also, we have to consider two sub-cases:

$*$ $Q(x, y) = p(xy)$, yielding $Q(x, y) = (xy - 1)(p_{11}xy - p_{-1,1})$;

$*$ $Q(x, y) = x^2 p\left(\dfrac{y}{x}\right)$, yielding $Q(x, y) = (y - x)(p_{-1,1}y - p_{1,-1}x)$.

(b) $Q = f_1 f_2$, *where f_1 is of degree 0 with respect to x.* Let us prove in fact f_1 is of degree 1 in y. Assume for a while it has degree 2 in y. Then $Q(x, y) = f_1(y)f_2(x)$ and both f_1 and f_2 have degree 2, so that exists α, β, γ, where

$$a(x) = \alpha f_2(x), \quad b(x) = \beta f_2(x), \quad c(x) = \gamma f_2(x).$$

Hence $\alpha > 0, \gamma > 0$ and f_2 has positive coefficients. But $b(x)$ has one negative coefficient, so that $\beta < 0$, which yields $p_{-10} = p_{10} = 0$ and $f_2(x) = \lambda x, \lambda > 0$, contrary to the hypothesis.

Hence f_1 has degree 1 in y and we can write $f_1(y) = y + \beta, \beta \neq 0$, so that

$$Q(x, y) = (y + \beta)\left(a(x)y + \frac{c(x)}{\beta}\right),$$

which yields

$$b(x) = \beta a(x) + \frac{c(x)}{\beta}.$$

As $b(x)$ has one strictly negative coefficient, one must have $\beta < 0$ together with $p_{10} = p_{-1,0} = 0$. It follows also that

$$p_{11} = p_{1,-1} = p_{-1,1} = p_{-1,-1} = 0.$$

Similarly, the situation obtained by exchanging the roles of x and y would give

$$Q(x, y) = (x + \widetilde{\beta}) \left(\widetilde{a}(y)x + \frac{\widetilde{c}(y)}{\widetilde{\beta}} \right).$$

The proof of lemma 2.3.2 is terminated.   ■

2.3.1 Branches of the Algebraic Functions on the Unit Circle

Let $\mathcal{P}$ be the simplex

$$\mathcal{P} = \left\{ (p_{-1,-1}, \ldots, p_{11}) : \sum_{i,j} p_{ij} = 1, \; p_{ij} \geq 0 \right\}.$$

The points of this simplex are the parameters of a random walk inside the quarter-plane. In the simplex $\mathcal{P}$, we shall often consider the set $\mathcal{A}$ of points for which $M_y = 0$. This set subdivides $\mathcal{P}$ into two convex sets $\mathcal{A}_+$ and $\mathcal{A}_-$, for which $M_y > 0$ and $M_y < 0$ respectively.

Definition 2.3.3   *A random walk is said to be simple if only p_{10}, p_{01}, $p_{-1,0}$, $p_{0,-1}$ are not equal to 0.*

Lemma 2.3.4   *Let the random walk be non singular and $M_y \neq 0$. Then the algebraic function $Y(x)$ has on Γ two branches, denoted by $Y_0(x)$ and $Y_1(x)$, such that $Y_0(1) < Y_1(1)$. There are two possible cases.*

(i) *If $M_y > 0$, then $|Y_0(x)| < 1$ and $|Y_1(x)| \geq 1.\forall |x| = 1$ and the equality takes place only for $x = 1$, where $Y_1(1) = 1$. Moreover, $Y_0(x)$ (resp. $Y_1(x)$), $x \in \Gamma$, is a real analytic curve, situated inside (resp. outside) the unit circle Γ, for $x \neq 1$.*

(ii) *If $M_y < 0$, then $|Y_0(x)| \leq 1$, for $|x| = 1$ and the equality takes place only for $x = 1$, where $Y_0(1) = 1$. Moreover, $|Y_1(x)| > 1$, $\forall x$.*

When $M_x \neq 0$, similar properties hold for the algebraic function $X(y)$, the two branches of which are denoted respectively by $X_0(y)$ and $X_1(y)$.   ■

Proof.   First, we shall state some simple assertions:

(i) Let $P(x,y) = \sum p_{ij} x^i y^j$. Then the equation

$$P(x,y) = 1, \quad |x| = |y| = 1,$$

cannot hold for non singular random walks, except for $x = y = 1$. This follows from elementary considerations on the sum of complex numbers.

(i) The equation $P(1,y) = 1$ has two roots, 1 and $\dfrac{c(1)}{a(1)}$ (see 2.3.1), and the two following situations can arise:

(a) if $M_y < 0$, then $Y_0(1) = 1$, $Y_1(1) = \dfrac{c(1)}{a(1)}$;

(b) if $M_y > 0$, then $0 < Y_0(1) = \dfrac{c(1)}{a(1)} < 1$, $Y_1(1) = 1$.

Take $|x| = 1$ and assume $a(x) \neq 0$. Then the roots in y of $a(x)y^2 + b(x)y + c(x)$ are continuous with respect to $a(x)$, $b(x)$ and $c(x)$. Consequently, the image $Y(\Gamma)$ consists of no more than two connected components. Thus, by (a), this image can intersect Γ only at the point $y = 1$. The case corresponding to the points $\{|x| = 1, a(x) = 0\}$ is obtained by continuity.

Assume for example $M_y < 0$. (The argument hereafter would be the same for $M_y > 0$, just replacing $\mathcal{D}$ by $\mathbb{C} \setminus \overline{\mathcal{D}}$). It follows now from (ii) that, if $Y(\Gamma)$ is connected, then $Y(\Gamma) \cap \mathcal{D} = \emptyset$. Hence proving $Y(\Gamma) \cap \mathcal{D} \neq \emptyset$ will show at once that $Y(\Gamma)$ has exactly two connected components: one of which, denoted by $Y_0(\Gamma)$, belongs to $\overline{\mathcal{D}}$ and the other one, $Y_1(\Gamma)$, belongs to $\mathbb{C} \setminus \overline{\mathcal{D}}$. For $x \in \Gamma$, $Y_0(x)$ and $Y_1(x)$ are simple closed analytic curves, and this amounts to say that $Y(x)$ has no branch point on Γ. This is really the case, because at a branch point $Y(x)$ would take only one value, which is impossible due to (a) and (b).

It remains to prove the existence of points x, $|x| = 1$, and y, $|y| < 1$, such that $P(x,y) = 1$. It suffices to check the particular case $x = -1$. To this end, we shall use a powerful continuity argument.

Any point $\rho \in \mathcal{A}_-$ can be connected, along a continuous path $\ell \subset \mathcal{A}_-$, with some point $\rho_0 \in \mathcal{A}_-$ corresponding to a *simple* random walk (see definition above). Moreover ℓ can be chosen to ensure the continuity of the two roots of $P(-1, y) = 1$ with respect to the parameters of $\mathcal{A}_-$, provided that one does not cross the hyperplane

$$a(-1) = p_{11} - p_{10} + p_{-1,1} = 0.$$

Next one checks easily that, for any *simple* random walk,

$$-1 < Y_0(-1) < 1,$$

whence it follows that $Y_0(-1) \subset \mathcal{D}$ for *any* random walk and the proof of the lemma is concluded. ∎

2.3.2 Branch Points

Our concern is now to locate the branch points of the algebraic function, which, in particular, define the genus of the Riemann surface. Starting from *simple* random walks, we will deduce general properties by using continuity arguments with respect to the parameters (i.e. the points of $\mathcal{P}$), as in lemma 2.3.4.

Lemma 2.3.5   *Consider a simple random walk satisfying $M_y \neq 0$. Then $Y(x)$ has two branch points x_1, x_2 inside $\mathcal{D}$ and two branch points x_3, x_4 outside $\mathcal{D}$. All these four branch points are positive and equal to (with an obvious notation)*

$$x_{1,2} = \frac{1 \pm 2\sqrt{p_{01}p_{0,-1}} - \sqrt{1 \pm 4\sqrt{p_{01}p_{0,-1}} + 4p_{01}p_{0,-1} - 4p_{10}p_{-1,0}}}{2p_{10}},$$

$$x_{3,4} = \frac{1 \pm 2\sqrt{p_{01}p_{0,-1}} + \sqrt{1 \pm 4\sqrt{p_{01}p_{0,-1}} + 4p_{01}p_{0,-1} - 4p_{10}p_{-1,0}}}{2p_{10}}.$$

When $M_y = 0$, $M_x \neq 0$, there are still four positive branch points, which satisfy

$$0 < x_1 < x_2 = 1 < x_3 < x_4, \quad if \quad M_x < 0,$$
$$0 < x_1 < x_2 < 1 = x_3 < x_4, \quad if \quad M_x > 0.$$

∎

Proof.   Factorizing the discriminant $\mathcal{D}_x$ of the equation

$$p_{01}y^2 + y(xp_{10} + \frac{p_{-1,0}}{x} - 1) + p_{0,-1} = 0,$$

we get the two following relationships which define the branch points, after having set $g(x) \overset{def}{=} p_{10}x + p_{-1,0}\frac{1}{x}$, namely

$$g(x) = 1 + 2\sqrt{p_{01}p_{0,-1}}, \tag{2.3.3}$$

$$g(x) = 1 - 2\sqrt{p_{01}p_{0,-1}}. \tag{2.3.4}$$

Since $g(1) = p_{10} + p_{-1,0} < 1 - 2\sqrt{p_{01}p_{0,-1}}$ and $g''(x) > 0$, the equation (2.3.3) has two positive zeros, one is bigger and the other is less than one. The same is true for equation (2.3.4). Other cases can be treated quite similarly.   ∎

The next two preliminary lemmas will be used in the proof of the subsequent lemmas 2.3.8, 2.3.9 and 2.3.10.

Lemma 2.3.6   *The point $x = -1$ is a branch point of $Y(x)$ if $M_y = 0$ and either of the two conditions is satisfied*

$$\begin{cases} p_{10} = p_{01} = p_{-1,0} = p_{0,-1} = 0, \\ p_{11} = p_{1,-1} = p_{-1,1} = p_{-1,-1} = p_{10} = p_{-10} = 0. \end{cases}$$

∎

Proof.

$$D(-1) = (1 + p_{10} - p_{00} + p_{-1,0})^2 - 4(p_{11} - p_{01} + p_{-1,1})(p_{1,-1} - p_{0,-1} + p_{-1,-1})$$
$$\geq (a(1) + c(1))^2 - 4a(1)c(1) = (a(1) - c(1))^2 = M_y^2.$$

Thus $D(-1) = 0$ yields first $D(1) = 0$, i.e. the point $x = 1$ is a branch point of $Y(x)$. Moreover the equalities $D(-1) = D(1) = 0$ imply that

(i) $b(-1) = -b(1) = a(1) + c(1)$,

and, either

(ii) $\begin{cases} a(-1) = a(1), \\ c(-1) = c(1), \end{cases}$

or

(iii) $\begin{cases} a(-1) = -a(1), \\ c(-1) = -c(1). \end{cases}$

It is readily been that *(i) + (ii)* [resp. *(i) + (iii)*] is tantamount to the first [resp. second] condition stated in the lemma. The proof of lemma 2.3.6 is concluded. ∎

Lemma 2.3.7   *The hypersurface $p_{10}^2 - 4p_{11}p_{1,-1} = 0$ subdivides $\mathcal{A}_-$ into two pathwise connected domains $\mathcal{A}_{-+}$ and $\mathcal{A}_{--}$, where $p_{10}^2 > 4p_{11}p_{1,-1}$ or $p_{10}^2 < 4p_{11}p_{1,-1}$ correspondingly. An analogous property is true for the hypersurface $p_{-1,0}^2 - 4p_{-1,1}p_{-1,-1} = 0$ and the corresponding domains $\mathcal{A}_+$, $\mathcal{A}_{++}$, and $\mathcal{A}_{+-}$.* ∎

Proof.   The proof is easy and left to the reader. ∎

Now we shall formulate all information about the general case in the next three lemmas.

Lemma 2.3.8   *For all non singular r.w. such that $M_y \neq 0$, $Y(x)$ has two branch points x_1 and x_2 (resp. x_3 and x_4) inside (resp. outside) the unit circle. All these branch points lie on the real line.*

∗ *For the pair (x_3, x_4), the following classification holds:*

1. *If $p_{10} > 2\sqrt{p_{11}p_{1,-1}}$, then x_3 and x_4 are positive;*

2. *If $p_{10} = 2\sqrt{p_{11}p_{1,-1}}$, then one point is infinite and the other is positive;*

3. *If $p_{10} < 2\sqrt{p_{11}p_{1,-1}}$, then one point is positive and the other is negative.*

* *Similarly, for the pair* (x_1, x_2),

  4. *if* $p_{-1,0} > 2\sqrt{p_{-1,1}p_{-1,-1}}$, *then* x_1 *and* x_2 *are positive;*

  5. *if* $p_{-1,0} = 2\sqrt{p_{-1,1}p_{-1,-1}}$, *then one point is 0 and the second is positive;*

  6. *if* $p_{-1,0} < 2\sqrt{p_{-1,1}p_{-1,-1}}$, *then one point is positive and the other is negative.*

This lemma is true also for $X(y)$, *up to a proper symmetric change of the parameters.* ∎

Proof.   Again let us connect an arbitrary point $\rho \in \mathcal{A}_{-+}$, by a continuous path $\ell \subset \mathcal{A}_{-+}$, with some point $\rho_0 \in \mathcal{A}_{-+}$ corresponding to a non degenerate *simple* random walk.

Due to lemma 2.3.5, all zeros of the discriminant at the point ρ_0 are real and mutually different. Along ℓ they will be real and different if a *liaison* does not occur (note that the zeros of the discriminant are two by two complex conjugate). But this discriminant is equal to

$$D(x) = [b(x) - 2\sqrt{a(x)c(x)}][b(x) + 2\sqrt{a(x)c(x)}].$$

At ρ_0, the points x_3 and x_4 are respectively zeros of the two equations

$$b(x) - 2\sqrt{a(x)c(x)} = 0 \text{ or } b(x) + 2\sqrt{a(x)c(x)} = 0.$$

Along ℓ, these zeros remain zeros of these different equations and cannot become equal since otherwise

$$b(x) - 2\sqrt{a(x)c(x)} = b(x) + 2\sqrt{a(x)c(x)} = 0,$$

which is possible only for $x = 0$ or $x = \infty$, but these cases have been considered above.

Let us now consider $\mathcal{A}_{--}$ and a point $\rho_0 \in \mathcal{P}$, corresponding to a r.w. for which $p_{ij} \neq 0$, if, and only if, $|ij| = 1$. The zeros of $D(x)$ in this case satisfy

$$x^2 = \frac{b \pm \sqrt{b^2 - 64p_{11}p_{1,-1}p_{-1,1}p_{-1,-1}}}{8p_{11}p_{1,-1}},$$

where

$$b = 1 - 4p_{11}p_{-1,-1} - 4p_{-1,1}p_{1,-1}.$$

One can prove (see lemma 2.3.2) that, if either $p_{11} \neq p_{-1,-1}$ or $p_{1,-1} \neq p_{-1,1}$, then

$$b^2 - 64p_{11}p_{-1,1}p_{1,-1}p_{-1,-1} > 0.$$

Clearly, $b \pm \sqrt{b^2 - 64p_{11}p_{1,-1}p_{-1,1}p_{-1,-1}} > 0$. Thus, two zeros of $D(x)$ are positive and two are negative. The fact that two of them lie inside and two other ones outside the unit circle follows from the fact (proven above) that, in an arbitrary neighborhood, there are points corresponding to non-degenerate r.w. for which this is true. The rest of the proof is similar, with the simplification that x_3 and x_4 cannot become equal, since one of them is > 1 and the other < -1. The sets $\mathcal{A}_{+-}$ and $\mathcal{A}_{++}$ could be handled in the same way. The proof of lemma 2.3.8 is terminated. ∎

Lemma 2.3.9   *For all non singular r.w. such that $M_y = 0$, one of the branch points of $Y(x)$ is equal to 1. In addition,*

* *if $M_x < 0$, then two other branch points have a modulus bigger than 1 and the remaining one has a modulus less than 1;*

* *if $M_x > 0$, then two branch points are less than 1 and the modulus of the remaining one is bigger than 1.*

Furthermore, the positivity conditions are the same as in lemma 2.3.8. ∎

Proof.   Consider the convex set $\mathcal{A}_0$ of the points ρ corresponding to a non-generate random walk with $M_y = 0$. The hypersurface defined by $M_x = 0$ divides this set into two convex sets for which, respectively, $M_x > 0$ and $M_x < 0$. The point $x = 1$ is always a branch point of $Y(x)$ and for a *simple* random walk the assertion in question holds, by lemma 2.3.5. To conclude the proof, we proceed exactly as in lemma 2.3.8. ∎

Lemma 2.3.10   *For all non singular random walks, $\mathbf{S}$ has genus 0 if, and only if, one of the following relations takes place:*

$$M_x = M_y = 0, \tag{2.3.5}$$
$$p_{10} = p_{11} = p_{01} = 0, \tag{2.3.6}$$
$$p_{10} = p_{1,-1} = p_{0,-1} = 0, \tag{2.3.7}$$
$$p_{-1,0} = p_{-1,-1} = p_{0,-1} = 0, \tag{2.3.8}$$
$$p_{01} = p_{-1,0} = p_{-1,1} = 0. \tag{2.3.9}$$

The singular random walks (c) and (e) on figure 2.3.1 correspond also to the genus 0 case and (2.3.5) implies

$$x_2 = x_3 = 1 \quad and \quad y_2 = y_3 = 1.$$

∎

Proof. $\mathbf{S}$ has genus 0 if, and only if, the discriminant

$$D(x) = b^2(x) - 4a(x)c(x) = d_4 x^4 + d_3 x^3 + d_2 x^2 + d_1 x + d_0$$

of the equation $Q(x,y) = a(x)y^2 + b(x)y + c(x) = 0$ has a multiple zero, possibly infinite, where we have put

$$
\begin{cases}
d_0 &= p_{-1,0}^2 - 4p_{-1,1}p_{-1,-1}\,, \\
d_1 &= 2p_{-1,0}(p_{00} - 1) - 4(p_{-1,1}p_{0,-1} + p_{01}p_{-1,-1})\,, \\
d_2 &= (p_{00} - 1)^2 + 2p_{10}p_{-1,0} - 4[p_{11}p_{-1,-1} + p_{1,-1}p_{-1,1} + p_{01}p_{0,-1}]\,, \\
d_3 &= 2p_{10}(p_{00} - 1) - 4(p_{11}p_{0,-1} + p_{01}p_{1,-1})\,, \\
d_4 &= p_{10}^2 - 4p_{11}p_{1,-1}\,.
\end{cases}
$$

The point of the proof is that multiple roots can occur only at $x = 0$, $x = 1$ or $x = \infty$. To see this, we use the continuity of the roots with respect to the points of $\mathcal{P}$ and the results of lemmas 2.3.8 and 2.3.9. Indeed, as $\mathcal{A}$ is convex, any arbitrary point ρ can be connected by a direct line ℓ with a point $\rho_0 \in \mathcal{A}$, which corresponds to a non-degenerate *simple* random walk. The zeros of the discriminant continuously depend on the point of ℓ and cannot intersect Γ: this yields the first assertion of the lemma, when $M_y < 0$. The case $M_y > 0$ can be treated in a similar way.

* There is a double root at $x = 1$ if, and only if, $M_x = M_y = 0$. To see this, write

$$
\begin{cases}
D(1) &= (a(1) - c(1))^2 = M_y^2 = 0, \\
D'(1) &= -4a(1)[b'(1) + a'(1) + c'(1)] = -4a(1)M_x = 0.
\end{cases}
$$

* There is a multiple zero at ∞ if, and only if, $d_3 = d_4 = 0$. It is easy to see that the system of equations $b_3 = b_4 = 0$, with respect to the p_{ij}'s, has admissible solutions if, and only if, (2.3.5) or (2.3.6) or the singular case on figure 2.3.1c take place.

* Similarly $D(x)$ has a multiple root at 0 if, and only if, (2.3.7) or (2.3.8) or the singular case on figure 2.3.1e holds.

The proof of lemma 2.3.10 is concluded. ∎

2.4 Galois Automorphisms and the Group of the Random Walk

Choose an arbitrary non singular random walk. This means among other things that $Q(x, y)$, considered as a polynomial in y over the field $\mathbb{C}(x)$ of rational functions of x, is irreducible over this field.

Consider also the vector space over $\mathbb{C}(x)$, generated by the constant 1 and one zero $y(x)$ of Q. This vector space is a field, which is the extension of order 2 of $\mathbb{C}(x)$ and will be denoted in the sequel by $\mathbb{C}(x)[y(x)]$. Each element of $\mathbb{C}(x)[y(x)]$ can be written in a unique way as $u(x) + v(x)y(x)$, where u and v are elements of $\mathbb{C}(x)$. Then it is quite natural to identify $\mathbb{C}(x)[y(x)]$ with the quotient field $\mathbb{C}(x)[T]/Q(x, T)$. Similarly, exchanging x and y, one can define $\mathbb{C}(y)[x(y)]$ and $\mathbb{C}(y)[T]/Q(T, y)$.

Let $\mathbb{C}(x,y)$ be the field of rational functions in (x,y) over $\mathbb{C}$. Since Q is assumed to be irreducible in the general case, the quotient ring of $\mathbb{C}(x,y)$ is in fact a field, which will be denoted by $\mathbb{C}_Q(x,y)$.

Proposition 2.4.1 *The fields $\mathbb{C}(x)[T]/Q(x,T)$ and $\mathbb{C}(y)[T]/Q(T,y)$ are isomorphic to $\mathbb{C}_Q(x,y)$*

Proof.   Noting that $\forall p \in \mathbb{C}_Q(x,y)$, $\exists$ a unique pair $u(x), v(x)$ of elements of $\mathbb{C}(x)$, such that

$$p = u(x) + v(x)y, \quad \mathrm{mod}\,Q.$$

Now the stated isomorphism is simply given by the mapping

$$i_x \; : \; \{u(x) + v(x)T\} \to \{u(x) + v(x)y\}\,,$$

where the brackets stand for the adequate equivalence classes. Exchanging the roles of x and y, there exists an isomorphism

$$i_y \; : \; \{u(y) + v(y)T\} \to \{u(y) + v(y)x\}$$

where $u(y)$ and $v(y)$ are elements of $\mathbb{C}(y)$. ∎

It is worth summarizing some of the above results by means of the following chain of statements, where the symbol $\cong$ means " isomorphic to " :

$$\mathbb{C}_Q(x,y) \cong \mathbb{C}(x)[T]/Q(x,T) \cong \mathbb{C}(y)[T]/Q(T,y) \cong \mathbb{C}(x)[y(x)] \cong \mathbb{C}(y)[x(y)]\,.$$

The Galois group of $\mathbb{C}(x)[y(x)]$ (resp. $\mathbb{C}(y)[x(y)]$) is cyclic of order 2 and its generic element will be denoted by ξ (resp. η), so that

$$\xi(u(x)) = u(x), \qquad \forall u \in \mathbb{C}(x), \tag{2.4.1}$$

$$\xi(y(x)) = \frac{c(x)}{y(x)a(x)} = -\frac{b(x)}{a(x)} - y(x), \tag{2.4.1'}$$

$$\eta(w(y)) = w(y), \qquad \forall w \in \mathbb{C}(y), \tag{2.4.2}$$

$$\eta(x(y)) = \frac{\widetilde{c}(y)}{x(y)\widetilde{a}(y)} = -\frac{\widetilde{b}(y)}{\widetilde{a}(y)} - x(y), \tag{2.4.2'}$$

It is important to remember that ξ [resp. η] *permutes the two roots* in y [resp. x] of $Q(x,y) = 0$.

In fact the above proposition means that there are two automorphisms $\widetilde{\xi}$ and $\widetilde{\eta}$ of $\mathbb{C}_Q(x,y)$, given respectively by

$$\widetilde{\xi} = i_x \circ \xi \circ i_x^{-1} \quad \text{and} \quad \widetilde{\eta} = i_y \circ \eta \circ i_y^{-1}, \tag{2.4.3}$$

where i_x and i_y have been defined in proposition 2.4.1. Up to a slight abuse in the notation, we will write

$$\begin{aligned}
\widetilde{\xi}(f(x,y)) &= f(x,\xi(y)) \mod Q, \quad \forall f \in \mathbb{C}(x,y), \\
\widetilde{\eta}(g(x,y)) &= g(\eta(x),y) \mod Q, \quad \forall g \in \mathbb{C}(x,y).
\end{aligned}$$

2.4.1 Construction of the Automorphisms $\hat{\xi}$ and $\hat{\eta}$ on S

We use theorem 2.1.13, taking $X = \mathbb{P}_x$, $Y = S$, $n = 2$. Then, by definition 2.1.10, propositions 2.1.12, 2.1.15, $\mathcal{M}(\mathbb{P}_x) = \mathbb{C}(x)$ and $\mathcal{M}(S)$ is an extension of degree 2 of $\mathbb{C}(x)$, so that it is *Galois*. Theorem 2.1.13 states the existence of an automorphism $\hat{\xi}$, belonging to $\mathrm{Aut}(S/\mathbb{P}_x)$, which is a cyclic group of order 2, isomorphic to $G(\mathbb{C}(x)[y(x)]/\mathbb{C}(x))$. According to paragraph 2.2.3 and equation (2.2.8), $\mathbb{C}(x)$ can be embedded into $\mathcal{M}(S)$ by the correspondence

$$\hat{f} : S \to \mathcal{M}(S), \quad \text{with } \hat{f}(s) \stackrel{\text{def}}{=} f(x(s)), \quad \forall f \in \mathbb{C}(x),$$

so that, to avoid superfluous notation, we will identify $\mathcal{M}(S)$ with $\mathbb{C}(x)[y(x)]$. Thus ξ, which is a generator of $G(\mathbb{C}(x)[y(x)]/\mathbb{C}(x))$ can be viewed as acting on $\mathcal{M}(S)$ and $\hat{\xi}$ is given by the relation

$$f \circ \hat{\xi} = \xi^{-1} \circ f, \quad \forall f \in \mathcal{M}(S).$$

Similarly,

$$f \circ \hat{\eta} = \eta^{-1} \circ f, \quad \forall f \in \mathcal{M}(S).$$

Letting I_d be the identity operator, the following set of relations hold:

$$h_x \circ \hat{\xi} = h_x \qquad \hat{\xi}^2 = I_d , \tag{2.4.4}$$

$$h_y \circ \hat{\eta} = h_y \qquad \hat{\eta}^2 = I_d . \tag{2.4.5}$$

The points s and $\hat{\xi}(s)$ [resp. $\hat{\eta}(s)$] have the same projections onto $\mathbb{P}_x$ [resp. $\mathbb{P}_y$]. In particular, the branch points of h_x [resp. h_y] are the fixed points of $\hat{\xi}$ [resp. $\hat{\eta}$], which read $h^{-1}(x_i)$ [resp. $h^{-1}(y_i)$], $i = 1, \dots, 4$.

Definition 2.4.2   *The group of the random walk is the group $\mathcal{H}$ of automorphisms of $\mathbb{C}_Q(x,y)$ generated by $\widetilde{\xi}$ and $\widetilde{\eta}$ (given in equation (2.4.3)), and it depends only on the transition probabilities $p_{ij}, |i|, |j| \leq 1$. This group is isomorphic to a subgroup of automorphisms of S generated by $\hat{\xi}$ and $\hat{\eta}$.*

Notation   Whenever no ambiguity arises, the following conventions will take place throughout the rest of this monograph:

- the composition of automorphisms will be written as an ordinary product;

- as far as global properties are concerned, one will simply write ξ, η, δ, without the symbols $\widetilde{\ }$ or $\hat{\ }$.

The next lemma gives the structure of all groups with generators ξ and η, such that $\xi^2 = \eta^2 = I_d$.

Lemma 2.4.3   *Define*

$$\delta = \eta\xi. \tag{2.4.6}$$

Then $\mathcal{H}$ has a normal cyclic subgroup $\mathcal{H}_0 = \{\delta^n, n \in \mathbb{Z}\}$, which is finite or infinite, and $\mathcal{H}/\mathcal{H}_0$ is a cyclic group of order 2. ∎

Proof.   We want to show that $x\delta^n x^{-1} \in \mathcal{H}_0$, for any $x \in \mathcal{H}$ and any integer n. But, since $\eta = \eta^{-1}$ and $\xi = \xi^{-1}$, this is a direct consequence of the following general result [43]. *Let G be a group and H a subgroup of G of index 2; then H is normal in G.* Recalling that the *index* is the number of distinct *left* cosets, i.e. subsets of G of the form xH, for all $x \in G$ (here x takes only the values ξ or η). ∎

2.5 Reduction of the Main Equation to the Riemann Torus

Here the Riemann surface is supposed to have genus 1 (which excludes non singular random walks) and we return to equations (2.2.8) and (2.2.9), first studying the geometric properties of the domains $h_x^{-1}(\mathcal{D})$, $h_y^{-1}(\mathcal{D})$ and of their boundaries $h_x^{-1}(\Gamma)$, $h_y^{-1}(\Gamma)$. For the sake of conciseness we shall only consider here the cases

1. $M_y < 0, \quad M_x < 0,$

2. $M_y < 0, \quad M_x > 0,$

3. $M_y < 0, \quad M_x = 0.$

The remaining patterns of interest are obtained by exchanging x and y in the above inequalities. It is also worth noting that the case $M_x > 0, M_y > 0$, although yielding a non-ergodic situation, can topologically be handled by means of analogous arguments.

Since **S** has genus 1, $h_x^{-1}(\mathcal{D})$ is connected. To see this, it suffices to check that $h_x^{-1}(\mathcal{D})$ is *arcwise* connected. We have seen in section 2.3.2 that there are two branch points x_1 and x_2 inside $\mathcal{D}$. Choose two arbitrary points u, v in $h_x^{-1}(\mathcal{D})$ on **S** and draw two arcs $[h_x(u), x_1]$ and $[x_1, h_x(v)]$ in $\mathcal{D}$. The arcwise connectivity follows now from proposition 2.1.5 and the fact that x_1 corresponds to the unique point s_1 on **S**.

Of course, analogous properties hold for $h_y^{-1}(\mathcal{D})$ and $h_x^{-1}(\mathcal{D}) \cap h_y^{-1}(\mathcal{D}) \neq \emptyset$, since, by lemma 2.3.4, $Y(x)$ and $X(y)$ can simultaneously take on values inside $\mathcal{D}$. We will denote by Γ_0 and Γ_1, respectively, the connected components of the boundary $h_x^{-1}(\Gamma)$ of $h_x^{-1}(\mathcal{D})$, such that

$$\begin{cases} \Gamma_0 \subset h_x^{-1}(\Gamma) \cap \{s : |y(s)| \leq 1\} = \{|x(s)| = 1\} \cap \{|y(s)| \leq 1\}, \\ \Gamma_1 \subset h_x^{-1}(\Gamma) \cap \{s : |y(s)| \geq 1\} = \{|x(s)| = 1\} \cap \{|y(s)| \geq 1\}. \end{cases}$$

Since by hypothesis $M_y \neq 0$, the results of section 2.3 assert that Γ_0 and Γ_1 are closed analytic curves without self-intersections. The curves $\widetilde{\Gamma_0}$ and $\widetilde{\Gamma_1}$ are defined in an analogous way. For example, $\widetilde{\Gamma_0} \subset h_y^{-1}(\Gamma)$ and, on it, $|y(s)| = 1$, $|x(s)| \leq 1$.

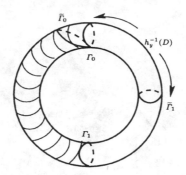

Fig. 2.5.1.

Let $M_x \neq 0$. The fact that Γ_0 and Γ_1 are homotopically equivalent follows from the construction of the Riemann surface $\mathbf{S}$, after taking into account that, inside and outside the unit circle Γ in the complex plane, there are exactly two branch points. This last property shows, in addition, that the *homology class* of Γ_0 and Γ_1 is one of the normal homology bases on the torus (see [72]), the same being true for $\widetilde{\Gamma_0}$ and $\widetilde{\Gamma_1}$.

Assume now for a while that e.g. Γ_0 and $\widetilde{\Gamma_0}$ belong to different homology classes, so that they would intersect. But, for $M_y < 0$, it has been shown in lemma 2.3.4 that $\Gamma_0 \cap \widetilde{\Gamma_0} = \emptyset$. [Clearly, if one would have supposed $M_y > 0$, by the same argument, one could have written $\Gamma_1 \cap \widetilde{\Gamma_1}$]. Hence, by transitivity, Γ_0, Γ_1, $\widetilde{\Gamma_0}$, $\widetilde{\Gamma_1}$ are defined by the same homology class.

The case $M_x = 0$ can be obtained by continuity, as mentioned in the next lemma, where the situation is completely summarized.

Lemma 2.5.1 *Define s_0 such that $x(s_0) = y(s_0) = 1$. The three following possibilities exist.*

Case 1 (Fig. 2.5.1)
$$M_y < 0, \quad M_x < 0.$$

Then $\Gamma_1 \cap \widetilde{\Gamma_1} = \emptyset$, and $\Gamma_0 \cap \widetilde{\Gamma_0} = h_x^{-1}(\Gamma) \cap h_y^{-1}(\Gamma)$ consists of the single point s_0.

Case 2 (Fig. 2.5.2)
$$M_y < 0, \quad M_x > 0.$$

Then $\Gamma_0 \cap \widetilde{\Gamma_1} = \{s_0\}$.

Case 3 (Fig. 2.5.3)
$$M_y < 0, \quad M_x = 0.$$

Then Γ_0 and Γ_1 do not intersect and one can define $\widetilde{\Gamma_0} = h_x^{-1}(\overline{\mathcal{D}}) \cap h_y^{-1}(\Gamma)$

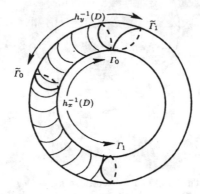

Fig. 2.5.2.

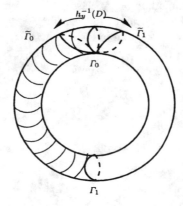

Fig. 2.5.3.

and $\widetilde{\Gamma_1} = h_y^{-1}(\Gamma) \cap (\mathbf{S} \backslash h_x^{-1}(\mathcal{D}))$, so that any pair of curves belonging to the triple $(\Gamma_0, \widetilde{\Gamma_0}, \widetilde{\Gamma_1})$ intersect at the point s_0 only.

The proof in Case 3 is just the same, by letting in Case 1 $\varepsilon \to 0$ in $h_y^{-1}(\Gamma_\varepsilon)$, where $\Gamma_\varepsilon = \{x : |x| = 1 + \varepsilon\}$. ∎

As already pointed out above, the case $M_x \geq 0$, $M_y \geq 0$, $M_x + M_y > 0$, is not considered, since then the random walk is transient (see [29]), but the construction is clearly of the same nature. It is a worthwhile exercise (left to the reader) to draw the pictures corresponding to the random walks **B** and **C**, introduced in lemma 2.3.2.

Thus the functions $\hat{\pi}$ and $\hat{\tilde{\pi}}$ are analytic in $h_x^{-1}(\mathcal{D}) \cap h_y^{-1}(\mathcal{D})$. They admit a mero-morphic continuation to the domains $h_y^{-1}(\mathcal{D})$ and $h_x^{-1}(\mathcal{D})$ respectively, provided that in these domains they are defined by

$$\hat{\pi} = -\frac{\hat{\tilde{q}}\,\hat{\tilde{\pi}} + \hat{q}_0 \pi_{00}}{\hat{q}}, \qquad \hat{\tilde{\pi}} = -\frac{\hat{q}\,\hat{\pi} + \hat{q}_0 \pi_{00}}{\hat{\tilde{q}}}.$$

In $h_y^{-1}(\mathcal{D})$ [resp. $h_x^{-1}(\mathcal{D})$], the only possible poles of $\hat{\pi}$ [resp. $\hat{\tilde{\pi}}$] are the zeros of $\hat{q}$ [resp. $\hat{\tilde{q}}$], since $\hat{q}$ [resp. $\hat{\tilde{q}}$] is a meromorphic function on $\mathbf{S}$.

In other words, $\hat{\pi}$ and $\hat{\tilde{\pi}}$ are meromorphic in $h_x^{-1}(\mathcal{D}) \cup h_y^{-1}(\mathcal{D})$ and in this region equation (2.2.9) still holds. One can now reformulate lemma 2.2.1, for non-degenerate random walks, as a problem set on the Riemann surface **S**.

Problem   Find two meromorphic functions $\hat{\pi}$ and $\hat{\tilde{\pi}}$ in $h_x^{-1}(\mathcal{D})$ and $h_y^{-1}(\mathcal{D})$ respectively, such that the following relations hold:

$$\begin{aligned} \hat{\pi}(s) &= \hat{\pi}(\hat{\xi}(s)), \\ \hat{\tilde{\pi}}(s) &= \hat{\tilde{\pi}}(\hat{\eta}(s)), \end{aligned}$$

when $(s, \hat{\xi}(s))$ and $(s, \hat{\eta}(s))$ belong to $h_x^{-1}(\mathcal{D}) \cup h_y^{-1}(\mathcal{D})$. Now it is easy to prove lemma 2.2.1. Assume for instance that Case 1 prevails, i.e. $M_x < 0$, $M_y < 0$. Looking at the figure 2.5.1, one can see that $\hat{\pi}$ is also defined in a neighborhood of Γ_0 on S. Now, one *lower* back $\hat{\pi}$ onto $\mathbb{C}$. From the very definition of all unknown functions and the probabilistic context, π must have no pole on Γ and, consequently, in some neighborhood of Γ. But π is a function of the single variable x. Since the function $h_x(s)$ has no branch point and is locally invertible in a neighborhood of Γ_0, it follows also that $\pi(x)$, $x \in \mathcal{D}_{1+\varepsilon}$, has no branch point and is holomorphic in this region.

Cases 2 and 3 of lemma 2.5.1 would give rise to slightly more complicated argu-ments and they will be considered later, in a more general context in chapters 3 and 5.

Remark 2.5.2 *The case of genus* 0, *corresponding to the Riemann sphere can be treated analogously, as was partially done in [56]. Indeed, it will be completely worked out in the complex plane in chapter 6.*

3. Analytic Continuation of the Unknown Functions in the Genus 1 Case

In section 2.5, we have shown, using the Riemann surface $\mathbf{S}$, that the functions π and $\widetilde{\pi}$ could be analytically continued to $\mathcal{D}_{1+\varepsilon}$. In this chapter, we shall propose other methods of analytic continuation, which in a sense are more effective, since they allow in fact a continuation to the whole complex plane. We consider only parameter values ensuring the Riemann surface $\mathbf{S}$ to be of genus 1, according to the study made in section 2.3.

3.1 Lifting the Fundamental Equation onto the Universal Covering

We will use some properties, which are classical only for readers acquainted with the theory of Riemann surfaces, and can be found either in Forster [34] or in Hurwitz and Courant [38]. Since $\mathbf{S}$ is of genus 1, its covering manifold is the complex plane $\mathbb{C}$ and its universal covering has the form $(\mathbb{C}, \lambda)$, where, by proposition 2.1.11, λ is *Galois* and $\mathrm{Aut}(\mathbb{C}/\mathbf{S})$ is isomorphic to the fundamental group of $\mathbf{S}$, which we denote by $\mathbf{T}$. In addition, it is known that

$$\mathbf{T} = \mathbb{Z}\omega_1 \oplus \mathbb{Z}\omega_2,$$

where ω_1 and ω_2 are two complex numbers, linearly independent on $\mathcal{R}$, acting as generators of $\mathbf{T}$. Thus, any $t \in \mathbf{T}$ can be written in the form

$$t = m_1\omega_1 + m_2\omega_2, \quad m_1, m_2 \in \mathbb{Z},$$

and, $\forall \tau \in \mathrm{Aut}(\mathbb{C}/\mathbf{S})$, there exists a unique $t \in \mathbf{T}$, with

$$\tau(\omega) = \omega + t, \quad \omega \in \mathbb{C}.$$

In the next section, we will calculate these periods explicitly (they are, in general, defined up to some unimodular transformation, see e.g. [38]).

Along lines quite similar to those of section 2.2.4, any meromorphic function f defined on $\mathbf{S}$ can be lifted onto the universal covering $\mathbb{C}$ by the mapping

$$f^* = f \circ \lambda. \tag{3.1.1}$$

Clearly, any such f^* is meromorphic and satisfies the functional equation

$$f^*(\omega) = f^*(\omega + m_1\omega_1 + m_2\omega_2), \quad \forall m_1, m_2 \in \mathbb{Z}. \tag{3.1.2}$$

Let us recall that a meromorphic function on $\mathbb{C}$ satisfying (3.1.2) is called an *elliptic function* with periods ω_1, ω_2. Thus the field $\mathbb{C}(\mathbf{S})$ of meromorphic functions on $\mathbf{S}$ is isomorphic, by (3.1.1), to the field of elliptic functions with the corresponding periods.

In addition, the period parallelogram

$$\Pi \stackrel{\text{def}}{=} \{\alpha_1\omega_1 + \alpha_2\omega_2 : 0 \le \alpha_i < 1\},$$

provided that its opposite sides are identified, is isomorphic to $\mathbf{S}$. Then $\lambda[0, \omega_1]$ and $\lambda[0, \omega_2]$ constitute an homology basis on the torus.

Let us fix the notation so that $\lambda\{[0, \omega_1]\}$ be homologous to Γ_0 and, therefore, also to the curves $\Gamma_1, \widetilde{\Gamma}_0, \widetilde{\Gamma}_1$ and consider the domain

$$\Delta \stackrel{\text{def}}{=} h_x^{-1}(\mathcal{D}) \cup h_y^{-1}(\mathcal{D}) \subset \mathbf{S}.$$

Its inverse image $\lambda^{-1}(\Delta)$ belongs to $\mathbb{C}$ and consists of a denumerable number of curvilinear strips, which differ from each other by a translation of vector ω_2. Any such strip is simply connected and invariant by translation of amplitude ω_1. Indeed, Δ is generated by a set of closed curves, say $\mathcal{L}(t), 0 \le t \le 1$, homotopic to Γ_0, as was shown in section 2.5 (see e.g. figure 2.5.1). Each curve $\mathcal{L}(t)$ has in Π a preimage homologous to $[0, \omega_1[$.

Denoting by Δ^* the connected component of $\lambda^{-1}(\Delta)$ having a non empty intersection with Π, one can use the mapping λ to lift π and $\widetilde{\pi}$ (which are meromorphic in Δ) onto Δ^*, just setting

$$\pi^* = \hat{\pi} \circ \lambda, \quad \widetilde{\pi}^* = \hat{\widetilde{\pi}} \circ \lambda.$$

In addition, as quoted in section 2.2.4, the coefficients $\hat{q}, \hat{\widetilde{q}}, \hat{q}_0$, are meromorphic functions on $\mathbf{S}$, which in turn can be lifted onto the universal covering and satisfy equation (3.1.2).

Notation It is the right place to make here a notational convention: each function, whatever its domain of definition may be, will be represented by the same symbol with the adequate arguments, e.g. $\pi(s), \pi(\omega), \widetilde{q}(s), \widetilde{q}(\omega), q(x, y)$, etc.

Hence $\pi(\omega)$ and $\widetilde{\pi}(\omega)$ are well defined and meromorphic in Δ^*, where, from their single valuedness along Γ_0 and $\widetilde{\Gamma}_0$ respectively, they satisfy the equations

$$\pi(\omega + \omega_1) = \pi(\omega), \quad \widetilde{\pi}(\omega) = \widetilde{\pi}(\omega + \omega_1), \tag{3.1.3}$$

$$q(\omega)\pi(\omega) + \widetilde{q}(\omega)\widetilde{\pi}(\omega) + q_0(\omega)\pi_{00} = 0. \tag{3.1.4}$$

Note that it suffices to analyze equation (3.1.3) in the domain $\Delta^* \cap \Pi$.

3.1.1 Lifting of the Branch Points

Starting from the branch points x_i, y_i, previously introduced in section 2.3.2, let $s_i, \tilde{s}_i \in \mathbf{S}$ be such that

$$h_x(s_i) = x_i, \quad h_y(\tilde{s}_i) = y_i, \quad \forall i = 1, \ldots, 4.$$

Clearly, s_i and $\tilde{s}_i$ are uniquely defined by these relations and consequently there exist unique points

$$a_i \overset{\text{def}}{=} \lambda^{-1}(s_i) \cap \Pi, \quad b_i \overset{\text{def}}{=} \lambda^{-1}(\tilde{s}_i) \cap \Pi, \quad \forall i = 1, \ldots, 4.$$

Moreover,

- for $i = 1, 2$, the points a_i and b_i belong to $\Delta^* \cap \Pi$,   since $s_i, \tilde{s}_i \in \Delta$;

- for $i = 3, 4$, the points a_i and b_i belong to $\Pi \setminus \Delta^*$.

3.1.2 Lifting of the Automorphisms on the Universal Covering

The goal of this section is to compute explicitly the automorphisms on the universal covering.

Let $\hat{\alpha}$ be an arbitrary automorphism of $\mathbf{S}$. The mapping

$$\hat{\alpha} \circ \lambda : \mathbb{C} \to \mathbf{S}$$

is holomorphic, so that, choosing in definition 2.1.4

$$X = \mathbf{S}, \quad Y = Z = \mathbb{C}, \quad h = \lambda, \quad f = \hat{\alpha} \circ \lambda,$$

we get a lifting of f with respect to λ, which will be denoted by α^* and, clearly, does satisfy the relation $f = \lambda \circ \alpha^*$. Since $(\mathbb{C}, \lambda)$ is a cover of $\mathbf{S}$ without branch point, it follows from proposition 2.1.6 that α^* is also holomorphic. Taking now

$$\hat{\alpha} \equiv \hat{\xi},$$

where $\hat{\xi}$ has been constructed in section 2.4.1, we get the corresponding automorphism ξ^* on $\mathbb{C}$, which satisfies the relation

$$\lambda \circ \xi^* = \hat{\xi} \circ \lambda. \tag{3.1.5}$$

Since

$$\lambda(a_i) = s_i, \quad \hat{\xi}(s_i) = s_i, \quad \forall i = 1, \ldots, 4,$$

it follows that

$$\lambda(a_i) = \lambda(\xi^*(a_i)).$$

Hence, there exist integers $k_{1,i}, k_{2,i} \in \mathbb{Z}$ with

$$\xi^*(a_i) = a_i + k_{1,i}\omega_1 + k_{2,i}\omega_2, \quad \forall i = 1, \ldots, 4.$$

Still using proposition 2.1.6, we know that ξ^* is unique, provided that one of its values is fixed. It is always possible to make a translation to make a_1 a fixed point of ξ^*, so that

$$\xi^*(a_1) = a_1.$$

Equation (3.1.5) yields immediately

$$\hat{\xi}^2 \circ \lambda = \lambda \circ \xi^{*2},$$

whence, since the range of $\xi^{*2} - I_d$ belongs to $\mathbb{Z}\omega_1 \oplus \mathbb{Z}\omega_2$ and $\hat{\xi}^2 = I_d$,

$$\xi^{*2} - I_d = K,$$

where K is a constant. Upon applying this last relation to the fixed point a_i, we get indeed $K = 0$. Thus we have just proved that ξ^* is an automorphism satisfying

$$\xi^{*2} = I_d.$$

Obviously any automorphism ζ^* of the complex plane has the form

$$\zeta^*(\omega) = \alpha\omega + \beta, \quad \forall \omega \in \mathbb{C}.$$

Hence, discarding the trivial automorphism I_d, one obtains easily

$$\xi^*(\omega) = -\omega + 2a_1. \tag{3.1.6}$$

So one can write, $\forall i = 1, \ldots, 4$,

$$\xi^*(a_i) = -a_i + 2a_1 = a_i + k_{1,i}\omega_1 + k_{2,i}\omega_2,$$

and, using (3.1.6),

$$a_1 - a_i = \frac{k_{1,i}\omega_1 + k_{2,i}\omega_2}{2},$$

where $k_{1,i}, k_{2,i} \in \{-1, 0, 1\}$, since all a_i's are located in Π. Remembering (see beginning of section 3.1) that $\lambda([0, \omega_1])$ was chosen to be homologous to Γ_0 on $\mathbf{S}$, one concludes that the open curve $\lambda([a_1, a_2])$ is homologous to *one half* of Γ_0. Thus $|a_1 - a_2| = \dfrac{\omega_1}{2}$ and we will take

$$a_2 - a_1 = \frac{\omega_1}{2}. \tag{3.1.7}$$

Mutatis mutandis, one can prove along the same lines that it is possible to lift η, to obtain an automorphism η^* of $\mathbb{C}$, such that

$$\eta^*(\omega) = -\omega + 2b_1, \quad b_2 - b_1 = \frac{\omega_1}{2}. \tag{3.1.8}$$

Setting

$$\omega_3 = 2(b_1 - a_1), \tag{3.1.9}$$

we can write, in accordance with equation (2.4.6),

$$\delta^* = (\eta\xi)^* = \eta^*\xi^*,$$

so that

$$\delta^*(\omega) = \omega + \omega_3. \tag{3.1.10}$$

3.2 Analytic Continuation

Theorem 3.2.1 π and $\widetilde{\pi}$ can be continued as meromorphic functions to the whole universal covering, where they satisfy equations (3.1.3), (3.1.4), together with the relations

$$\pi(\omega) = \pi(\xi^*(\omega)), \quad \widetilde{\pi}(\omega) = \widetilde{\pi}(\eta^*(\omega)), \quad \forall \omega \in \mathbb{C}. \tag{3.2.1}$$

∎

Proof.   The region Δ^* is simply connected. Moreover, $\Delta^* \cap \Pi$ is bounded by the curves $\lambda^{-1}(\Gamma_1) \cap \Pi$ and $\lambda^{-1}(\widetilde{\Gamma}_1) \cap \Pi$. From the analysis made in sections 2.4 and 2.5, the following *automorphy* relationships on $\mathbf{S}$ hold:

$$\hat{\xi}(\Gamma_1) = \Gamma_0, \quad \hat{\delta}(\Gamma_1) = \hat{\eta}(\Gamma_0) \subset \overline{\Delta}, \quad \hat{\xi}(\widetilde{\Gamma}_0) \subset \overline{\Delta}, \tag{3.2.2}$$

using in particular the fact that η preserves x and $\Gamma_0 \subset \{s, |x(s)| \leq 1\}$.

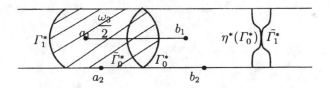

Fig. 3.2.1. The case $M_x < 0$, $M_y < 0$

In figure 3.2.1, the symbol " $*$ " has been used for all curves, to emphasize that we are on the universal covering, and δ^* is a translation *to the right* of vector $\omega_3 = 2(a_1 - b_1)$ in the system of coordinates (ω_2, ω_1), according to (3.1.10). The detailed computation of $\omega_1, \omega_2, \omega_3$ is carried out in section 3.3. Lifting now the relations in (3.2.2) onto the universal covering, one can write for instance

$$\xi^*(\lambda^{-1}(\Gamma_1) \cap \Pi) = \lambda^{-1}(\Gamma_0) \cap \Pi, \quad \eta^*(\lambda^{-1}(\widetilde{\Gamma}_1) \cap \Pi) = \lambda^{-1}(\widetilde{\Gamma}_0) \cap \Pi,$$

whence

$$\delta^*(\lambda^{-1}(\Gamma_1) \cap \Pi) \subset \overline{\Delta^*} \cap \Pi, \quad \delta^{*-1}(\lambda^{-1}(\widetilde{\Gamma}_1) \cap \Pi) \subset \overline{\Delta^*} \cap \Pi. \tag{3.2.3}$$

Thus the domain

$$\Delta^* \bigcup \delta^*(\Delta^*)$$

is in fact simply connected and it follows by induction that

$$\bigcup_{n=-\infty}^{\infty} \delta^{*n}(\Delta^*) \tag{3.2.4}$$

covers the whole complex plane $\mathbb{C}$. Moreover, since on $\mathbb{C}_x$ π is a function of $x, \forall |x| \leq 1$, we have

$$\begin{cases} \pi(\omega) = \pi(\xi^*(\omega)), & \forall(\omega, \xi^*(\omega)) \in \Delta^*, \\ \widetilde{\pi}(\omega) = \widetilde{\pi}(\eta^*(\omega)), & \forall(\omega, \eta^*(\omega)) \in \Delta^*, \end{cases}$$

and this equality, by the principle of analytic continuation, holds indeed for all $\omega \in \mathbb{C}$, so that (3.2.1) is proved.

Using now the fundamental equation (3.1.4) together with (3.2.1), we obtain after some easy manipulations the following system:

$$\begin{cases} q(\xi^*(\omega))\pi(\omega) + \widetilde{q}(\xi^*(\omega))\widetilde{\pi}(\xi^*(\omega)) + q_0(\xi^*(\omega))\pi_{00} = 0, & \forall(\omega, \xi(\omega)) \in \Delta^*, \\ q(\xi^*(\omega))\pi(\omega) + \widetilde{q}(\xi^*(\omega))\widetilde{\pi}(\delta^*(\omega)) + q_0(\xi^*(\omega))\pi_{00} = 0, & \forall(\omega, \delta^*(\omega)) \in \Delta^*. \end{cases}$$

Eliminating $\pi(w)$ in the above system, we obtain

$$\begin{aligned} \widetilde{\pi}(\delta^*(\omega)) = \widetilde{\pi}(\omega + \omega_3) &= \frac{q(\xi^*(\omega))\widetilde{q}(\omega)}{\widetilde{q}(\xi^*(\omega))q(\omega)}\widetilde{\pi}(\omega) \\ &+ \frac{q(\xi^*(\omega))}{\widetilde{q}(\xi^*(\omega))}\left[-\frac{q_0(\xi^*(\omega))}{q(\xi(\omega))} + \frac{q_0(\omega)}{q(\omega)}\right]\pi_{00}. \end{aligned} \qquad (3.2.5)$$

Equation (3.2.5) allows us to continue $\widetilde{\pi}(\omega)$, first to the domain $\delta^*(\Delta^*)$, and then, by induction, to the region introduced in (3.2.4). The proof of theorem 3.2.1 is concluded. ∎

The next step consists in making the analytic continuation of $\pi(x)$ and $\widetilde{\pi}(y)$ in their respective complex planes $\mathbb{C}_x$ and C_y. To this end, we first go through an intermediate step, namely the analytic continuation of π and $\widetilde{\pi}$ on the Riemann surface $\mathbf{S}$. This is the subject of the next theorem.

Theorem 3.2.2 *The functions π and $\widetilde{\pi}$ can be continued to the whole Riemann surface $\mathbf{S}$.* ∎

Proof.   Indeed, since the mapping $\lambda : \mathbb{C} \to \mathbf{S}$ is holomorphic, π and $\widetilde{\pi}$ (which by theorem 3.2.1 are known to be meromorphic functions on the universal covering $\mathbb{C}$) can be projected onto $\mathbf{S}$ also as meromorphic, but possibly multi-valued functions (since the torus $\mathbf{S}$ is not simply connected), according to the usual procedure, as follows.

Any path ℓ on $\mathbf{S}$, having its origin at $s \in \Delta$, can be lifted onto a path ℓ^*, which is unique provided that the origin s^* of ℓ^* is fixed in $\lambda^{-1}(s)$, since λ has no branch point: it suffices to impose $s^* \in \lambda^{-1}(s) \cap \Pi$. This choice of s and s^* yields two functions on $\mathbf{S}$, which coincide on Δ with π and $\widetilde{\pi}$, respectively, and are defined by

$$\pi \circ \lambda^{-1}(t), \quad \widetilde{\pi} \circ \lambda^{-1}(t), \quad \forall t \in \ell. \qquad (3.2.6)$$

From the monodromy theorem, the continuation along any curve homotopic to Γ_0 leads to the same branch. In order to obtain the other branches of the two above-mentioned functions, it suffices to make the analytic continuation along some curve non homotopic to Γ_0. The proof of the theorem is concluded.   ∎

The pragmatic problem mentioned above can be rephrased as follows: *starting with $\pi(x)$ defined for $|x| \leq 1$, we ask whether this function can be continued outside the unit disk.* The solution consists in projecting onto $\mathbb{C}_x$ and $\mathbb{C}_y$ respectively the functions defined on **S** by (3.2.6) .

Let us draw a cut in $\mathbb{C}_x$, along the real axis between the points x_3 and x_4. More exactly, this cut is

$$
\begin{cases}
[x_3, x_4], & \text{if } x_3 < x_4 \leq \infty, \\
[x_3, \infty] \cup [-\infty, x_4], & \text{if } x_4 < -1,
\end{cases}
$$

but in all cases we will write $[x_3, x_4]$.

Theorem 3.2.3   *Under the conditions of theorem 1.2.1, $\pi(x)$ is a meromorphic function in the complex plane $\mathbb{C}_x$ cut along $[x_3, x_4]$. A similar statement holds for $\widetilde{\pi}(y)$, with the corresponding cut $[y_3, y_4]$ in $\mathbb{C}_y$.*   ∎

Proof.   One can proceed exactly along the lines which were used above to project π from $\mathbb{C}$ onto **S**. Choose a path ℓ in $\mathbb{C}_x - \mathcal{D}$, not going through x_i, $i = 3, 4$. Since above any $u \in \ell$ there are exactly two points on **S**, $h_x^{-1}(\ell)$ consists of *two* curves and $h_x(s)$ is *locally* one to one $\forall s \neq s_i$, by proposition 2.1.6. Denoting one of these curves arbitrarily by $\mathcal{L}$, we have $h_x^{-1}(\ell) = \mathcal{L} \cup \hat{\xi}(\mathcal{L})$. On the other hand, from section 2.5 and the construction presented above, it becomes clear that
$$
\pi(s) = \pi(\hat{\xi}(s)), \quad \forall s \in \mathbf{S}.
$$

Thus, taking $\mathcal{L}$ or $\hat{\xi}(\mathcal{L})$ does not matter and this shows that the analytic continuation of π along any arbitrary path ℓ in $\mathbb{C}_x - \mathcal{D}$ is possible. Here again, the continued function is *a priori* multi-valued. To render it single-valued, it suffices to cut $\mathbb{C}_x$ along $[x_3, x_4]$. Indeed, any simple closed curve ℓ in $\mathbb{C}_x \setminus \{[x_3, x_4] \cup \overline{\mathcal{D}}\}$ is lifted onto **S** into a curve $\mathcal{L}$, homotopic either to a single point or to Γ_0 (depending wether or not $\mathcal{L}$ contains $[x_3, x_4]$ in its interior), but never to a latitude circle on the torus, which would correspond to the segment $[0, \omega_2]$ on the universal covering. Besides, we know that π is single-valued on such $\mathcal{L}$, since we remain on the same branch. The proof of theorem 3.2.3 is concluded.   ∎

Corollary 3.2.4   *Under the conditions of theorem 1.2.1, $\pi(x)$ and $\widetilde{\pi}(y)$ are holomorphic in the neighborhood of the unit circle, in their respective complex planes.*

The proof is obvious since $[x_3, x_4] \subset \mathbb{C}_x - \overline{\mathcal{D}}$.   ∎

Corollary 3.2.5   *The following conditions are equivalent:*

1. π is not rational;

2. $\widetilde{\pi}$ is not rational;

3. π (resp. $\widetilde{\pi}$) is not meromorphic in $\mathbb{C}_x$ (resp. $\mathbb{C}_y$);

4. x_3 (resp. y_3) is a non polar singularity of π (resp. $\widetilde{\pi}$).

Proof. $4 \rightarrow 3 \rightarrow 1$ is clear. To show $1 \rightarrow 2$, assume π is rational. Then, from the basic functional equation (2.2.4), we have

$$\widetilde{\pi}(y) = E(y) + X_0(y)F(y),$$

where E and F are rational functions of y. But $\widetilde{\pi}$ has to be holomorphic in the unit disk, so that, necessarily, $F(y) \equiv 0$, which says that $\widetilde{\pi}$ is then also rational. For the induction $1 \rightarrow 3 \rightarrow 4$, two cases have to be considered:

• $x_4 \neq \infty$. One sees easily, using analytic properties of the branches given in sections 2.3 and 5.3, that $x = \infty$ is either a regular point or a pole of π. Thus it remains to prove that, whenever x_3 is not a branch point of π, then this property holds also for x_4. But there exists no meromorphic function on $\mathbb{C}_x \backslash \{x_4\}$, having x_4 as a branch point.

• $x_4 = \infty$. Then, referring again to section 5.3, we have $y_4 \neq \infty$, provided the random walk is non singular, and we come to the preceding argument applied to the function $\widetilde{\pi}$.   ∎

3.3 More about Uniformization

Our purpose here is to get explicit representations for λ, ω_1 and ω_2, which will be used, in particular, at the end of chapter 5.

First we will recall some classical facts about the Weierstrass elliptic function $\wp(\omega; \omega_1, \omega_2)$, admitting the periods $m\omega_1 + n\omega_2$, $m, n \in \mathbb{Z}$, and usually denoted by $\wp(\omega)$. All this material can be found, for instance in [38] or [5]. $\wp(\omega)$ is defined by the series

$$\wp(\omega) = \frac{1}{\omega^2} + \sum_{(m,n) \neq (0,0)} \left[\frac{1}{(\omega - m\omega_1 - n\omega_2)^2} - \frac{1}{(m\omega_1 + n\omega_2)^2} \right],$$

which is uniformly and absolutely convergent for all $\omega \neq m\omega_1 + n\omega_2$, $m, n \in \mathbb{Z}$. At the points $m\omega_1 + n\omega_2$, $\wp(\omega)$ has poles of second order. The definition of $\wp(\omega; \omega_1, \omega_2)$ depends on the pair (ω_1, ω_2), up to a unimodular transform of the type

$$\omega_1' = \alpha\omega_1 + \beta\omega_2, \quad \omega_2' = \gamma\omega_1 + \delta\omega_2,$$

where $\alpha, \beta, \gamma, \delta$ are integers satisfying

$$\alpha\delta - \beta\gamma = \pm 1.$$

It is well known that $\wp(\omega)$ satisfies the differential equation

$$\wp'^2(\omega) = 4\wp^3(\omega) - g_2\wp(\omega) - g_3, \qquad (3.3.1)$$

where the so-called *invariants* g_2 *and* g_3 are given by

$$
\begin{cases}
g_2 = 60 \displaystyle\sum_{(m,n)\neq(0,0)} \frac{1}{(m\omega_1 + n\omega_2)^4}, \\[2ex]
g_3 = 140 \displaystyle\sum_{(m,n)\neq(0,0)} \frac{1}{(m\omega_1 + n\omega_2)^6}.
\end{cases}
\qquad (3.3.2)
$$

Moreover, any elliptic function with periods ω_1, ω_2, is a rational function of $\wp$ and of its derivative $\wp'$ and any *even* elliptic function is a rational function of the *sole* $\wp(u)$.

In section 2.3, we have introduced the polynomial

$$Q(x, y) = a(x)y^2 + b(x)y + c(x).$$

Setting temporarily, for the sake of shortness,

$$
\begin{cases}
D(x) = b^2(x) - 4a(x)c(x) = d_4 x^4 + d_3 x^3 + d_2 x^2 + d_1 x + d_0, \\
z = 2a(x)y + b(x),
\end{cases}
$$

with

$$
\begin{cases}
d_4 = p_{10}^2 - 4p_{11}p_{1,-1}, \\
d_3 = -[2p_{10}(1 - p_{00}) + 4p_{11}p_{0-1} + p_{01}p_{1-1}] < 0,
\end{cases}
$$

we know from chapter 2 that $D(x)$ has four real zeros x_i, $i = 1, \ldots, 4$. In addition x_2 and x_3 are always positive and we have

$$
\begin{cases}
-1 \leq x_1 < 1 \leq x_2 < x_3 < x_4, & \text{if } d_4 > 0, \\
x_4 < -1 \leq x_1 < 1 \leq x_2 < x_3, & \text{if } d_4 < 0, \\
x_4 = \infty, & \text{if } d_4 = 0.
\end{cases}
$$

Lemma 3.3.1   *When* **S** *is of genus 1, the algebraic curve* $Q(x, y) = 0$ *admits a uniformization given in terms of the Weierstrass $\wp$-function and its derivative $\wp'$ by the following formulas:*

(i) If $d_4 \neq 0$, *then* $D'(x_4) > 0$ *and*

$$
\begin{cases}
x(\omega) = x_4 + \dfrac{D'(x_4)}{\wp(\omega) - \frac{1}{6}D''(x_4)}, \\[3ex]
z(\omega) = \dfrac{D'(x_4)\wp'(\omega)}{2\left(\wp(\omega) - \frac{1}{6}D''(x_4)\right)^2};
\end{cases}
\qquad (3.3.3)
$$

(ii) If $d_4 = 0$, then $x_4 = \infty$ and

$$\begin{cases} x(\omega) = \dfrac{\wp(\omega) - \dfrac{d_2}{3}}{d_3}, \\[4mm] z(\omega) = -\dfrac{\wp'(\omega)}{2d_3}. \end{cases} \tag{3.3.4}$$

In both cases, $z(\omega)$ and $\wp'(\omega)$ have the same sign.

In the above formulas any x_i might have been chosen, but x_4 is in some sense the best candidate, since it depends on d_4 in a very direct way. ∎

Proof.   The algebraic curve $Q(x, y) = 0$ can be rewritten in the slightly simpler form

$$z^2 = D(x).$$

A classical way of uniformizing such curves, when D has degree 4 in x, consists into a reduction to the *Weierstrass domain*, in which case D would be of degree 3. To achieve this, it suffices to send one of the zeros of D, e.g. x_4, to infinity, using a fractional linear transformation.

Consider next the following Taylor's expansion

$$D(x) = (x - x_4)D'(x_4) + \frac{(x - x_4)^2}{2!}D''(x_4)$$
$$+ \frac{(x - x_4)^3}{3!}D^{(3)}(x_4) + \frac{(x - x_4)^4}{4!}D^{(4)}(x_4),$$

where $D^{(j)}(.)$ denotes the derivative of order $j \geq 3$, and let

$$u = \frac{D'(x_4)}{x - x_4}, \qquad v = \frac{2zD'(x_4)}{(x - x_4)^2}.$$

We have therefore

$$v^2 = 4u^3 + 2D''(x_4)u^2 + \frac{2u}{3}D^{(3)}(x_4)D'(x_4) + \frac{D^{(4)}(x_4)[D'(x_4)]^2}{6}.$$

Setting now

$$t = u + \frac{1}{6}D''(x_4),$$

yields the well known Weierstrass canonical form

$$v^2 = 4t^3 - g_2 t - g_3. \tag{3.3.5}$$

The case $d_4 = 0$ (and then $d_3 \neq 0$, since the degree of $D(x)$ must be ≥ 3 to have a curve of genus 1) is easier to handle, since it is then possible to eliminate the coefficient of degree 2 in D, by the linear change of variables

$$x = \frac{t - \dfrac{d_2}{3}}{d_3}, \quad z = -\frac{v}{2d_3},$$

which yields at once (3.3.5). Now the Weierstrass domain formed by the pairs (t, v) satisfying (3.3.5) is uniformized by setting directly, see [38],

$$t = \wp(\omega), \quad v = \wp'(\omega).$$

Hence, with these values for t, v, u, the asserted formulas (3.3.3) and (3.3.4) hold and the lemma is proved. ∎

It is well known that the periods ω_1, ω_2 of $\wp(\omega)$, which appear in the uniformization of (3.3.5), are uniquely defined in terms of g_2, g_3 by the formulas (3.3.3), since the ratio ω_2/ω_1 is assumed to be non real. Nonetheless these inversion formulas are not easy to manipulate numerically. Indeed they involve elliptic modular functions (see e.g. [5]). Thus, it seems useful to give a direct explicit form for the periods ω_1, ω_2 in terms of elliptic integrals.

Lemma 3.3.2   *One can choose ω_1, ω_2 so that ω_1 be purely imaginary and ω_2 be real. Such a choice is unique, up to signs. Hence we shall take $\omega_2 > 0$, $\operatorname{Im} \omega_1 > 0$. Moreover, ω_j are given by the following integrals, taken on intervals of the real axis:*

$$\omega_1 = 2i \int_{x_1}^{x_2} \frac{dx}{\sqrt{-D(x)}}, \quad \omega_2 = 2 \int_{x_2}^{x_3} \frac{dx}{\sqrt{D(x)}}, \tag{3.3.6}$$

where $D(x) \leq 0$, for $x_1 \leq x \leq x_2$, and $D(x) \geq 0$, for $x_2 \leq x \leq x_3$, and the radical in the integrand is taken to be positive. ∎

Proof.   Let us note first, that if $d_4 \neq 0$, then $D'(x_4) > 0$. Consequently, using the definition of $x(\omega)$ given in lemma 3.3.1, we show that, if $d_4 \neq 0$, then $\wp$ is a decreasing function of x, by (3.3.3), separately on the intervals $]-\infty, x_4[$ and $]x_4, +\infty[$. If $d_4 = 0$, then $\wp$ is a decreasing function of x on the whole real line $\mathbb{R}$, since $d_3 < 0$. The zeros of y, where $g(x) = 4x^3 - g_2 x - g_3$, are usually denoted by e_1, e_2, e_3 (see [5]). They are real and satisfy the following relationship

$$e_1 + e_2 + e_3 = 0, \quad e_1 > 0, \quad e_3 < 0, \quad e_1 > e_2 > e_3.$$

Let (ω_1, ω_2) be a pair of primitive periods. It is known (see [38]) that

$$\wp\left(\frac{\omega_2}{2}\right) = e_1, \quad \wp\left(\frac{\omega_1}{2}\right) = e_3, \quad \wp\left(\frac{\omega_1 + \omega_2}{2}\right) = e_2.$$

Consider the mapping $h = z \to \dfrac{z}{2}$ in the complex plane. This homothetie transforms the period parallelogram into another parallelogram, denoted by H. A property of the Weierstrass function $\wp(\omega)$ says that, when we describe H is the positive direction, starting from the point $\omega = 0$, $\wp(\omega)$ decreases from $+\infty$ to $-\infty$. This yields immediately, from (3.3.3) and (3.3.4),

$$x\left(\frac{\omega_2}{2}\right) = x_1, \quad x\left(\frac{\omega_1 + \omega_2}{2}\right) = x_2, \quad x\left(\frac{\omega_1}{2}\right) = x_3, \quad x(0) = x_4. \qquad (3.3.7)$$

Let now the elliptic integral

$$I(w) \stackrel{\text{def}}{=} \int_{-\infty}^{w} \frac{dt}{\sqrt{g(t)}}.$$

It is in fact more convenient to consider $I(w)$ on the Riemann surface $\mathbf{S}$, where the integrand becomes a meromorphic function. But then $I(w)$ is a multi-valued function. Fix a value of the integrand at some point a and consider two paths ℓ_1 and ℓ_2 on $\mathbf{S}$, such that the homologous paths L_1 and L_2, inside the period parallelogram in $\mathbb{C}$, join the points a to w. Then there exists integers m_1, m_2, such that

$$\int_{L_1} \frac{dt}{\sqrt{g(t)}} = \int_{L_2} \frac{dt}{\sqrt{g(t)}} + m_1 \omega_1 + m_2 \omega_2.$$

Taking $a = \infty$, $w = \wp(\omega; \omega_1, \omega_2)$, we get

$$\omega = \int_{\wp(\omega;\omega_1,\omega_2)} \frac{dt}{\sqrt{4t^3 - g_2 t - g_3}}, \qquad (3.3.8)$$

where ω, for any value of $\wp$, is defined modulo (ω_1, ω_2), noting that $\wp(\omega)$ is single-valued for all ω. It is worth noting that in (3.3.8) the radical is supposed to have a positive value for $\omega \in]0, \frac{\omega_2}{2}[$ and, in fact, the integral should be considered on the torus or, equivalently, in the period parallelogram. Thus at the point $\omega = \frac{\omega_2}{2}$, one passes from one sheet to the other, which means that, for $\omega \in]\frac{\omega_2}{2}, \omega_2[$, the radical is necessarily a negative real quantity.

Let us make now in (3.3.8) the change of variables

$$x = x_4 + \frac{D'(x_4)}{t - \frac{1}{6}D''(x_4)}, \quad \text{if } d_4 \neq 0,$$

$$x = \frac{t - \frac{d_2}{3}}{d_3}, \quad \text{if } d_4 = 0.$$

Then

$$\omega = \int_{x_4}^{x(\omega)} \frac{dx}{\sqrt{D(x)}}, \quad \text{mod } (\omega_1, \omega_2),$$

so that a particular pair of periods $(\widetilde{\omega}_1, \widetilde{\omega}_2)$ is given by

$$\widetilde{\omega}_1 = 2 \int_{x_4}^{x_3} \frac{dx}{\sqrt{D(x)}},$$

$$\widetilde{\omega}_2 = 2 \int_{x_4}^{x_1} \frac{dx}{\sqrt{D(x)}}.$$

In fact, $\widetilde{\omega}_1$ and $\widetilde{\omega}_2$ depend on the choice of the path of integration. In particular, it is always possible to take the path $\widehat{s_2 s_4}$ instead of $\widehat{s_1 s_4}$ on $\mathbf{S}$. This means that one can always take

$$\omega_2 = 2 \int_{x_2}^{x_3} \frac{dx}{\sqrt{D(x)}} > 0,$$

where the integration is taken over the real interval $[x_2, x_3]$. In addition, the following inequalities hold:

$$2 \int_{x_3}^{x_4} \frac{dx}{\sqrt{D(x)}} = \int_{L_2} \frac{dx}{\sqrt{D(x)}} = \int_{L_1} \frac{dx}{\sqrt{D(x)}} = 2 \int_{x_2}^{x_1} \frac{dx}{\sqrt{D(x)}},$$

up to the sign, which can be arbitrarily chosen. Since $D(x) \leq 0$, for $x \in [x_1, x_2]$, we get

$$\omega_1 = 2i \int_{x_1}^{x_2} \frac{dx}{\sqrt{-D(x)}},$$

as possible second primitive period, which is purely imaginary. Since the pair (ω_1, ω_2) is given up to a unimodular substitution, it appears that the above choice of (ω_1, ω_2) is unique, as asserted. The proof of lemma 3.3.2 is concluded. ∎

Lemma 3.3.3   *We have*

$$\omega_3 = 2 \int_{X(y_1)}^{x_1} \frac{dx}{\sqrt{D(x)}},$$

with

$$0 < \omega_3 < \omega_2.$$

∎

Proof.   We shall use the formulae (3.3.7), together with

$$x(a_1) = x_1, \quad a_1 = \frac{\omega_2}{2}, \quad x(b_1) = X(y_1),$$

where the three equations above simply depict the correspondance between the branch points x_i, y_i in the complex plane and their images $a_i, b_i, i = 1, 2$ on the universal covering.

The following properties of the function $\wp$ will be needed (see figure 3.3.1):

(i) $\wp$ is real on the sole intervals $[0, \omega_2[, [0, \omega_1[, [\frac{\omega_1}{2}, \frac{\omega_1}{2} + \omega_2[, [\frac{\omega_2}{2}, \frac{\omega_2}{2} + \omega_1[$;

(ii) $\wp$ decreases on $[0, \frac{\omega_2}{2}]$ from $+\infty$ to e_1, increases on $[\frac{\omega_2}{2}, \omega_2[$ from e_1 to $+\infty$. More generally, (see [5]), $\wp$ is real and monotone decreasing along the circuit drawn in figure 3.3.1.

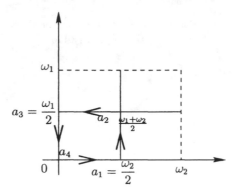

Fig. 3.3.1.

Thus, for $\omega \in \left[0, \frac{\omega_2}{2}\right]$, the function $x(\omega)$ (which is homographic of $\wp(\omega)$) is increasing and takes its values in the following intervals:

$$\begin{cases}]x_4, +\infty[\text{ and }]-\infty, x_1[, \text{ if } d_4 > 0; \\ [x_4, x_1], \text{ if } d_4 < 0; \\ ]-\infty, x_1[, \text{ if } d_4 = 0. \end{cases}$$

One will see later on in chapter 5 that $X(y_1) < x_1$, which implies first that b_1 is real and also that $b_1 \in]0, \omega_2[$, by the earlier mentioned properties of the function $\wp(\omega)$. Hence only two cases can take place:

$$\begin{cases} b_1 = c, \text{ where } c \in \left[0, \frac{\omega_2}{2}\right], \quad x(c) = X(y_1), \text{ or} \\ b_1 = \omega_2 - c. \end{cases}$$

To make the right choice, one must know the sign of $\wp'(b_1)$, since $\wp$ decreases on $\left[0, \frac{\omega_2}{2}\right]$ and increases on $\left[\frac{\omega_2}{2}, \omega_2\right]$. By lemma 3.3.1, $\wp'(b_1)$ and $z(b_1)$ have the same sign. Furthermore, from the very definition of z, we have

$$z(b_1) = 2a(X(y_1))y_1 + b(X(y_1)).$$

Referring again to section 2.3 and to chapter 5, the branch $Y_0(x)$ of the algebraic function $y(x)$ satisfies $y_1 = Y_0(X(y_1))$. Thus

$$\begin{aligned} z(b_1) &= 2a(X(y_1))Y_0(X(y_1)) + b(X(y_1)) \\ &= a(X(y_1))[Y_0(X(y_1)) - Y_1(X(y_1))] = \pm\sqrt{D(x(y_1))}, \end{aligned}$$

where we have used $Y_0(x) + Y_1(x) = \dfrac{-b(x)}{a(x)}$.

Taking now $x_1 - \varepsilon < x < x_1$, for some $\varepsilon > 0$, we will show in chapter 5 $Y_0(x) > Y_1(x)$. In addition, $a(x_1) > 0$. This last inequality can be proved by noting that, as $D(x) < 0$ on $]x_1, x_2[$, $a(x)$ has no zero on $[x_1, x_2]$. But $x_2 > 0$ implies $a(x_2) > 0$, so that $a(x_1) > 0$. Finally, the sign of the quantity

$$[Y_0(x) - Y_1(x)]a(x) = \pm\sqrt{D(x)}$$

does not change on the intervals

$$\begin{cases}]x_4, x_1[, \text{ if } x_4 < 0, \\ ]x_4, +\infty[\cup[-\infty, x_1[, \text{ if } x_4 > 0. \end{cases}$$

This shows that

$$[Y_0(X(y_1)) - Y_1(X(y_1))]a(X(y_1)) > 0,$$

which yields in turn

$$b_1 = \omega_2 - c.$$

Then

$$\omega_3 = 2(b_1 - a_1) = \omega_2 - 2c = 2\int_{x_4}^{x_1} \frac{dx}{\sqrt{D(x)}} + 2\int_{X(y_1)}^{x_4} \frac{dx}{\sqrt{D(x)}}.$$

Lemma 3.3.3 is proved.   ∎

Remark 3.3.4 *The calculation of the periods can also be achieved by using equations (3.3.2) and the modular functions [38].*

Remark 3.3.5 *All the derivations concerning the periods and the quantity ω_3 are connected with the fact that there exists a unique Abelian differential of the first kind, up to a multiplicative constant,*

$$\frac{dx}{2a(x)y + b(x)}.$$

This differential becomes $d\omega$ when lifted onto the universal covering and $\dfrac{dx}{\sqrt{D(x)}}$ when projected onto the complex plane $\mathbb{C}_x$.

4. The Case of a Finite Group

In section 2.4, the group $\mathcal{H}$ of the random walk was shown to be of even order $2n, n = 2, \ldots, \infty$. Throughout this chapter, n is supposed to be finite and the functions $q, \widetilde{q}, q_0$ are polynomials. In this case, we are able to characterize completely the solutions of the basic functional equation, and also to give necessary and sufficient conditions for these solutions to be rational or algebraic.

4.1 On the Conditions for $\mathcal{H}$ to be Finite

Finding the exact conditions in explicit form for n to be finite is a difficult question, which has many connections with some classical problems in algebraic geometry.

Recalling that $\mathcal{H}$ is generated by the elements ξ and η, we have the identity (homomorphism)

$$h(R(x, y)) \stackrel{\text{def}}{=} R(h(x), h(y)), \quad \forall h \in \mathcal{H}, \ \forall R \in \mathbb{C}_Q(x, y).$$

Clearly, 2 elements h_1, h_2 of $\mathcal{H}$ are identical if, and only if,

$$h_1(x) = h_2(x), \quad h_1(y) = h_2(y).$$

In addition, for any $R \in \mathbb{C}_Q(x, y)$, the following important equivalences hold:

$$\begin{cases} \xi(R) = R \iff & R \in \mathbb{C}(x), \\ \eta(R) = R \iff & R \in \mathbb{C}(y), \end{cases} \tag{4.1.1}$$

so that $\mathbb{C}(x)$ (resp. $\mathbb{C}(y)$ are the elements of $\mathbb{C}_Q(x, y)$ invariant with respect to ξ (resp. η). Indeed, let

$$\rho(x, y) = \frac{P_0(\dot{x}, y)}{P_1(x, y)}, \tag{4.1.2}$$

where $P_0(x, y), P_0(x, y)$ are polynomials in x, y. Since

$$Q(x, y) = a(x)y^2 + b(x)y + c(x),$$

(4.1.2) yields

$$\rho(x, y) = \frac{A_1(x)y + A_0(x)}{B_1(x)y + B_0(x)} \quad \mathrm{mod}\ Q(x, y).$$

When π is supposed to be invariant with respect to ξ, it follows that

$$\frac{A_1(x)y + A_0(x)}{B_1(x)y + B_0(x)} = \frac{A_1(x)y_\xi + A_0(x)}{B_1(x)y_\xi + B_0(x)} \quad \mathrm{mod}\ Q(x, y). \tag{4.1.3}$$

As $y \neq y_\xi$, we obtain from (4.1.3)

$$A_1(x)B_0(x) = A_0(x)B_1(x),$$

which yields immediately

$$\rho(x) = \begin{cases} \dfrac{A_0(x)}{B_0(x)}, & \text{if}\quad B_0 \neq 0, \\[2mm] \dfrac{A_1(x)}{B_1(x)} & \text{otherwise.} \end{cases} \tag{4.1.4}$$

Thus we have proved (4.1.1).
We shall now derive conditions for the order to be 4 or 6.

4.1.1 Explicit Conditions for Groups of Order 4 or 6

Lemma 4.1.1   *The order of $\mathcal{H}$ is 4 if, and only if,*

$$\begin{vmatrix} p_{11} & p_{10} & p_{1,-1} \\ p_{01} & p_{00} - 1 & p_{0,-1} \\ p_{-1,1} & p_{-1,0} & p_{-1,-1} \end{vmatrix} = 0. \tag{4.1.5}$$

∎

Proof.   Recalling that $\delta \overset{\text{def}}{=} \eta\xi$, the equality $\delta^2 = I_d$ writes

$$\xi\eta = \eta\xi, \quad \text{which is equivalent to} \quad \begin{cases} \xi\eta(x) = \eta(x), \\ \eta\xi(y) = \xi(y), \end{cases}$$

where we have used $\xi(x) = x$ and $\eta(y) = y$. Thus $\eta(x)$ [resp. $\xi(y)$] is left invariant by ξ [resp. η], so that

$$\eta(x) \in \mathbb{C}(x) \quad \text{and} \quad \xi(y) \in \mathbb{C}(y).$$

Hence, η and ξ are conformal automorphisms on $\mathbb{C}_x$ and $\mathbb{C}_y$ respectively, and are thus necessarily fractional linear transforms of the form (since $\xi^2 = \eta^2 = 1$),

$$\eta(x) = \frac{rx + s}{tx - r}, \quad \xi(y) = \frac{\widetilde{r}y + \widetilde{s}}{\widetilde{t}y - \widetilde{r}},$$

where all coefficients belong to $\mathbb{C}$. The following chain of equivalences holds:

$$\eta(x) = \frac{rx + s}{tx - r} \quad \Leftrightarrow \quad tx\eta(x) = r(x + \eta(x)) + s$$

$$\Leftrightarrow \quad 1, \; x + \eta(x), \; x\eta(x) \text{ are linearly dependent on } \mathbb{C}$$

$$\Leftrightarrow \quad 1, \; -\frac{\widetilde{b}(y)}{\widetilde{a}(y)}, \; \frac{\widetilde{c}(y)}{\widetilde{a}(y)} \text{ are linearly dependent on } \mathbb{C}$$

$$\Leftrightarrow \quad \widetilde{a}(y), \widetilde{b}(y), \widetilde{c}(y) \text{ are also linearly dependent on } \mathbb{C},$$

where we have used equation (2.3.1)

$$Q(x, y) = \widetilde{a}(y)x^2 + \widetilde{b}(y)x + \widetilde{c}(y)$$

and the results of section 2.4. The proof of lemma 4.1.1 is concluded.   ■

Lemma 4.1.2   *$\mathcal{H}$ has order 6 iff*

$$\begin{vmatrix} \Delta_{23} & \Delta_{33} & \Delta_{22} & \Delta_{32} \\ \Delta_{13} & -\Delta_{23} & \Delta_{12} & -\Delta_{22} \\ \Delta_{22} & \Delta_{32} & \Delta_{21} & \Delta_{31} \\ \Delta_{12} & -\Delta_{22} & \Delta_{11} & -\Delta_{21} \end{vmatrix} = 0, \qquad (4.1.6)$$

where Δ_{ij} denotes the determinant obtained from 4.1.5 by suppressing the i-th line and the j-th column.   ■

Proof.   In this case $(\xi\eta)^3 = I_d$, which is equivalent to

$$\eta\xi\eta = \xi\eta\xi. \qquad (4.1.7)$$

Applying (4.1.7) for instance to x, we get

$$\xi\eta(x) = \eta\xi\eta(x),$$

which shows that $\xi\eta(x)$ is invariant with respect to η and consequently is a rational function of y, remembering one is dealing with the field of rational functions. Similarly, $\eta\xi(y)$, invariant with respect to ξ, is a rational function of x. Hence (4.1.7) is plainly equivalent to

$$\begin{cases} \xi\eta(x) & = P(y), \\ \eta\xi(y) & = R(x), \end{cases}$$

where P and R are rational. Then

$$y = R(\xi\eta(x)) = R \circ P(y),$$

or, equivalently,

$$R \circ P = I_d, \qquad (4.1.8)$$

so that P and R are linear partial fractions. Thus (4.1.8) yields the relation

$$\xi(y) = \frac{p\eta(x) + q}{r\eta(x) + s}, \tag{4.1.9}$$

which imposes a linear dependence on $\mathbb{C}$ between the four elements $1, \xi(y), \eta(x), \xi(y)\eta(x)$. Moreover a simple but tedious computation leads to the following general relations, valid on the algebraic curve $Q(x, y) = 0$,

$$\begin{cases} \eta(x) &= -\dfrac{xu(y) + v(y)}{xw(y) + u(y)}, \\[2ex] \xi(y) &= -\dfrac{y\widetilde{u}(x) + \widetilde{v}(x)}{y\widetilde{w}(x) + \widetilde{u}(x)}, \end{cases} \tag{4.1.10}$$

with

$$\begin{aligned} u(y) &= \Delta_{22} + y\Delta_{23}, & \widetilde{u}(x) &= \Delta_{22} + x\Delta_{32}, \\ v(y) &= \Delta_{12} + y\Delta_{13}, & \widetilde{v}(x) &= \Delta_{21} + x\Delta_{31}, \\ w(y) &= \Delta_{32} + y\Delta_{33}, & \widetilde{w}(x) &= \Delta_{23} + x\Delta_{33}. \end{aligned}$$

Combining (4.1.9) and (4.1.10) leads to the condition (4.1.6) (the details of the calculus are omitted). The proof of lemma 4.1.2 is terminated. ∎

Examples where $\mathcal{H}$ is of order 4.

1. The product of 2 independent random walks inside the quarter plane, so that

$$\sum_{i,j} p_{ij} x^i y^j = p(x)\widetilde{p}(y).$$

2. The *simple* random walk where $\sum p_{ij} x^i y^j = p(x) + \widetilde{p}(y)$. Thus, in the interior of the quarter plane, $p_{ij} \neq 0$ if, and only if, i or j is zero.

3. The case (a) in figure 4.1.1, which can be viewed as a simple queueing network with parallel arrivals and internal transfers.

Examples where $\mathcal{H}$ is of order 6. Among these are the cases studied in [22], [6] and [31], which are represented in figure 4.1.1 (b), (c) respectively, where, as before, only jumps inside the quarter plane have been drawn.

4.1.2 The General Case

As quoted at the beginning of this chapter, the general question of finiteness of the group has indeed intimate links with problems encountered in geometry during the last century. Among them one can cite

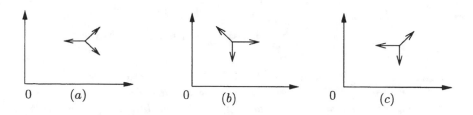

Fig. 4.1.1.

1. The problem of abelian integrals;

2. Poncelet's problem, pointed out in private discussions with the authors by L. Flatto.

We shall consider the first problem in more details. First, let us note that the group $\mathcal{H}$ is of order $2n$ if, and only if,

$$n\omega_3 = 0 \quad \mod (\omega_1, \omega_2),$$

or, since ω_3 is real,

$$n\omega_3 = 0 \quad \mod (\omega_2), \qquad (4.1.11)$$

where n is the minimal positive integer with this property.

It is convenient to recall hereafter some fundamental notions and theorems pertaining to compact Riemann surfaces. This material can be found e.g. in [72].

Let us take two arbitrary points u_1, u_2 on the universal covering, such that

$$u_2 - u_1 = \omega_3,$$

and let

$$P_1 = \lambda(u_1), \quad P_2 = \lambda(u_2),$$

their corresponding images on **S**. Then, from theorem 2.1.17, one can choose on **S** a unique (up to a multiplicative constant) abelian differential of the first kind $d\omega$, satisfying

$$\int_{P_1}^{P_2} d\omega = \omega_3. \qquad (4.1.12)$$

It follows from (4.1.11), (4.1.12) and Abel's theorem 2.1.18, that the divisor $\dfrac{P_1^n}{P_2^n}$ is *principal*. In other words, there exists a meromorphic function φ on **S**, with a unique zero (resp. pole) of n-th order at P_1 (resp. P_2). Clearly, the quantity

$$d(\log \varphi) = \frac{d\varphi}{\varphi}$$

is an abelian differential of the third kind, with poles of first order at P_1 and P_2.

Lemma 4.1.3   *Condition (4.1.11) holds if, and only if, there exists an abelian integral of the third kind, having logarithmic singularities at the points P_1, P_2 and represented as the logarithm of an algebraic function φ, which belongs to $\mathbb{C}(S)$.* ∎

Proof.   The necessary condition has been proved above. Let us now assume this integral does exist and denote it by $\omega_{P_1 P_2}$. Setting

$$\varphi = \exp \omega_{t_1 t_2},$$

and using again Abel's theorem, we come to (4.1.11). Moreover, the corresponding divisor is equal to $\dfrac{P_1^n}{P_2^n}$ and this n is the one we need. In other words, lemma 4.1.3 reduces the calculation of $\mathcal{H}$ to the following question: when is it possible to integrate a given abelian integral of the third kind in terms of logarithms ? Abel and Chebyshev have studied the latter problem. Its solution for integrals of the form

$$\int (z + C) \frac{dz}{\sqrt{\mathcal{D}(z)}},$$

where $\mathcal{D}(z)$ is the 4th order polynomial with real coefficients, was obtained by E.I. Zolotarev in 1874 [73]. Our situation reduces to integrals of exactly this kind. We shall see in section 4.2 that the group is finite if, and only if, the set $\mathbb{C}(x) \cap \mathbb{C}(y)$ is non trivial.

4.2 Rational Solutions

Let the group $\mathcal{H}$ be finite of order $2n$. We shall obtain, by strictly algebraic manipulations, rational solutions of the fundamental equation. These solutions have in general no probabilistic meaning, but they allow to reduce the problem to finding the solution of an homogeneous equation, i.e. when $q_0(x, y) = 0$. To find rational solutions, one can simply consider the main equation

$$-Q(x, y)\pi(x, y) = \pi(x)q(x, y) + \tilde{\pi}(y)\tilde{q}(x, y) + \pi_{00}q_0(x, y),$$

in $\mathbb{C}_Q(x, y)$. In other words, all calculations will be made *modulo Q*.

Notation   One will write $f_\alpha \stackrel{\text{def}}{=} \alpha(f)$, for all automorphisms α and all functions f belonging to $\mathbb{C}_Q(x, y)$.

Then we have

$$q\pi + \widetilde{q}\widetilde{\pi} + q_0\pi_{00} = 0, \tag{4.2.1}$$

$$\pi = \pi_\xi, \tag{4.2.2}$$

$$\widetilde{\pi} = \widetilde{\pi}_\eta \tag{4.2.3}$$

Applying η to (4.2.1), we get

$$\frac{q_\eta}{\widetilde{q}_\eta}\pi_\eta + \widetilde{\pi} + \frac{(q_0)_\eta}{\widetilde{q}_\eta}\pi_{00} = 0. \tag{4.2.4}$$

Eliminating now $\widetilde{\pi}$ from (4.2.1) and (4.2.4) yields

$$\frac{q_\eta}{\widetilde{q}_\eta}\pi_\eta - \frac{q}{\widetilde{q}}\pi + \left(\frac{(q_0)_\eta}{\widetilde{q}_\eta} - \frac{q_0}{\widetilde{q}}\right)\pi_{00} = 0,$$

or, since $\pi_\eta = \pi_{\eta\xi} = \pi_\delta$,

$$\pi_\delta - f\pi = \psi, \tag{4.2.5}$$

where

$$\begin{cases} \varphi = \dfrac{q}{\widetilde{q}}, \quad f = \dfrac{\varphi}{\varphi_\eta}, \quad \widetilde{f} = \dfrac{\varphi_\xi}{\varphi} \\[2mm] r = \dfrac{\pi_{00}q_0}{\widetilde{q}}, \quad \widetilde{r} = \dfrac{\pi_{00}q_0}{q}, \\[2mm] \psi = \dfrac{r - r_\eta}{\varphi_\eta}, \quad \widetilde{\psi} = \dfrac{\widetilde{r} - \widetilde{r}_\xi}{\varphi_\xi}, \\[2mm] \widetilde{\delta} \overset{\text{def}}{=} \delta^{-1} = \xi\eta. \end{cases} \tag{4.2.6}$$

Upon applying now δ repeatedly in (4.2.5), π satisfies the following system:

$$\begin{cases} \pi_\delta - f\pi & = \psi, \\ \pi_{\delta^2} - f_\delta\pi_\delta & = \psi_\delta, \\ \dots \\ \pi_{\delta^n} - f_{\delta^{n-1}}\pi_{\delta^{n-1}} & = \psi_{\delta^{n-1}}. \end{cases} \tag{4.2.7}$$

$\mathcal{H}$ being of order $2n$, we have $\delta^n = I_d$ and the system (4.2.7) is closed. Indeed, referring to section 2.4, since ex hypothesis π is sought in $\mathbb{C}_Q(x, y)$, we have

$$\pi_{\delta^n} = \pi.$$

Of course analogous relationships could be written for $\widetilde{\pi}$, using the automorphism $\widetilde{\delta}$ and the functions $\widetilde{f}, \widetilde{\psi}$. There are two deeply different situations, which will be analyzed in the next subsections. For this purpose, it is necessary to introduce some definitions, taken from algebra, with the notation employed in Lang [43].

Let us denote by $\mathbb{C}_\delta(x, y)$, the subfield formed by the elements of $\mathbb{C}_Q(x, y)$ which are invariant with respect to δ. Recalling that n is the order of the cyclic group $\mathcal{H}_0 = \{\delta^k, \ k \geq 0\}$ introduced in lemma 2.4.3, one can define, for any $f \in \mathbb{C}_Q(x, y)$, respectively a *trace* and a *norm* as follows:

$$
\begin{cases}
Tr(f) & \overset{\text{def}}{=} Tr^{\mathbb{C}_Q}_{\mathbb{C}_\delta}(f) \overset{\text{def}}{=} \sum_{\alpha \in \mathcal{H}_0} f_\alpha = f + f_\delta + \cdots + f_{\delta^{n-1}}, \\
N(f) & \overset{\text{def}}{=} N^{\mathbb{C}_Q}_{\mathbb{C}_\delta}(f) \overset{\text{def}}{=} \prod_{\alpha \in \mathcal{H}_0} f_\alpha = f f_\delta \cdots f_{\delta^{n-1}}.
\end{cases}
$$

It is worth noting that $Tr(f)$ and $N(f)$ are elements of $\mathbb{C}_\delta(x, y)$, invariant with respect to δ.

4.2.1 The Case $N(f) \neq 1$

Under the assumption

$$
N(f) \neq 1, \tag{4.2.8}
$$

the linear sytem (4.2.7) yields directly a rational solution ρ, with

$$
\rho = \frac{\sum_{i=0}^{n-1} \psi_{\delta^i} \prod_{k=i+1}^{n-1} f_{\delta^k}}{1 - N(f)}, \tag{4.2.9}
$$

with the usual convention that the empty product is equal to 1.

Similarly, under the condition

$$
N(\widetilde{f}) \neq 1, \tag{4.2.10}
$$

where $N(\widetilde{f})$ is computed with the automorphism $\widetilde{\delta}$, we obtain

$$
\widetilde{\rho} = \frac{\sum_{i=0}^{n-1} \widetilde{\psi}_{\delta^i} \prod_{k=i+1}^{n-1} \widetilde{f}_{\delta^k}}{1 - N(\widetilde{f})}. \tag{4.2.11}
$$

As the reader might have guessed, conditions (4.2.8) and (4.2.10) are in fact equivalent. To prove this property, it suffices to show that

$$
N(f) N(\widetilde{f}) = 1. \tag{4.2.12}
$$

Since, by (4.2.6),

$$
N(f) = \frac{N(\varphi)}{N(\varphi_\eta)} \quad \text{and} \quad N(\widetilde{f}) = \frac{N(\varphi_\xi)}{N(\varphi)},
$$

one must simply check $N(\varphi_\xi) = N(\varphi_\eta)$. But

$$
N(\varphi_\xi) = [N(\varphi)]_\xi = [N(\varphi)]_\eta = N(\varphi_\eta),
$$

where the intermediate equality is easy to derive. Hence (4.2.12) is proved.

Theorem 4.2.1   *Let the order of $\mathcal{H}$ be $2n$ and assume condition (4.2.8) or (4.2.10) holds. Then the system formed by (4.2.1), (4.2.2), (4.2.3) admits a unique rational solution, ρ, $\widetilde{\rho}$, given by (4.2.9) and (4.2.11) respectively.*   ■

Proof.   We have just proved uniqueness. To establish existence, one has only to show the equality $\rho = \rho_\xi$. Using the relations

$$f f_\eta = 1, \quad \frac{\varphi}{\varphi_\eta} = f, \quad \xi\delta^{n-i} = \widetilde{\delta}^{i+1}\eta,$$

we get

$$\begin{cases} \psi_{\xi\delta^{n-i}} = \psi_{\widetilde{\delta}^{i+1}\eta} = -\dfrac{\psi_{\widetilde{\delta}^{i+1}}}{f_{\widetilde{\delta}^{i+1}}}, \\[2ex] f_{\xi\delta^{n-i+1}} = (f_{\widetilde{\delta}^i})^{-1} = (f_{\delta^{n-i}})^{-1}, \end{cases}$$

whence

$$\rho_\xi = \frac{\left(\displaystyle\sum_{i=1}^{n} \psi_{\delta^{n-i}} \prod_{k=2}^{i} f_{\delta^{n-k+1}}\right)_\xi}{1 - \displaystyle\prod_{i=0}^{n-1} \frac{1}{f_{\delta^i}}} = \frac{-\displaystyle\sum_{i=1}^{n} \frac{\psi_{\widetilde{\delta}^{i+1}}}{f_{\widetilde{\delta}^{i+1}}} \prod_{k=2}^{i} \frac{1}{f_{\widetilde{\delta}^k}}}{1 - \displaystyle\prod_{i=0}^{n-1} \frac{1}{f_{\delta^i}}} = \rho. \qquad (4.2.13)$$

The proof of theorem 4.2.1 is concluded.   ■

Remark 4.2.2   *We have obtained the unique rational solution, which is non-zero if, and only if, $q_0 \neq 0$. Moreover whenever this solution has poles inside the unit circle, one can infer that the generating functions of the stationary probabilities cannot be rational.*

4.2.2 The Case $N(f) = 1$

Before proceeding to the detailed analysis, we quote in the forthcoming remark three properties, useful in the applications to avoid intricate computations, and can be derived along the lines which produced (4.2.12).

Remark 4.2.3   *The following equivalences hold:*

$$\begin{cases} N(f) = 1 \Longleftrightarrow \\ N(\varphi) \in \mathbb{C}(x) \cap \mathbb{C}(y) \Longleftrightarrow \\ N(\varphi) = [N(\varphi)]_\xi = [N(\varphi)]_\eta. \end{cases}$$

In section 4.2.1, it was shown that, whenever condition (4.2.8) takes place,

$$\pi = w + \rho, \qquad (4.2.14)$$

where $\rho \in \mathbb{C}(x)$ is given by theorem 4.2.1 and w satisfies the homogeneous equation

$$w_\delta - fw = 0. \tag{4.2.15}$$

Now, whenever

$$N(f) = 1,$$

it will be proved that

$$\pi = cu,$$

where $c \in \mathbb{C}(x)$ is rational and u satisfies the equation

$$u_\delta - u = c_\delta \psi.$$

En passant we shall also find all possible rational solutions.

Notation   Let F be an arbitrary field, h an automorphism of F. Then F_h will denote the subfield of elements of F which are invariant with respect to h.

Lemma 4.2.4   *Let F_h, F be two fields such that*

  (i) F is a finite Galois extension of F_h;

  (ii) The Galois group $G(F/F_h)$, (i.e. the set of automorphisms of F leaving F_h invariant – see section 2.1.3) is cyclic and generated by h.

Then, for any φ, $\psi \in F$, such that

$$\begin{cases} N_{F_h}^F(\varphi) & \overset{def}{=} \prod_{i=0}^{n-1} \varphi_{h^i} = 1, \\ Tr_{F_h}^F(\psi) & \overset{def}{=} \sum_{i=0}^{n-1} \psi_{h^i} = 0, \end{cases}$$

there exist a, $b \in F$, satisfying respectively

$$\begin{cases} \varphi = \dfrac{a}{a_h}, \\ \psi = b - b_h. \end{cases}$$

■

Proof.   The result follows directly from the multiplicative and additive forms of Hilbert's theorem 90, which we quote now for the sake of completeness (the proofs can be found e.g [43]).

Theorem 4.2.5 (Hilbert's theorem 90) *Let $K \subset L$ two arbitrary fields such that the Galois group $G(L/K)$ be cyclic of degree n. Let σ a generator of G. Let $\beta \in L$.*

(**Multiplicative form**) *The norm* $N_K^L(\beta)$ *is equal to* 1 *if, and only if, there exists an element* $a \neq 0$ *in* L *such that* $\beta = a/a_\sigma$.

(**Additive form**) *The trace* $Tr_K^L(\beta)$ *is equal to* 0 *if, and only if, there exists an element* b *in* L *such that* $\beta = b - b_\sigma$. ∎

Here we shall choose $L = F$ and $K = F_h$. The proof of the multiplicative form of Hilbert's theorem ensures, in a constructive manner, the existence of $\theta \in F$, such that

$$a = \theta + \varphi\theta_h + \varphi\varphi_h\theta_{h^2} + \cdots + \varphi\varphi_h \cdots \varphi_{h^{n-2}}\theta_{h^{n-1}} \neq 0.$$

Such an a is admissible.

Similarly, for any $\gamma \in F$ with $Tr_{F_h}^F(\gamma) \neq 0$, an admissible b is given by

$$b = \frac{1}{Tr_{F_h}^F(\gamma)}[\psi\gamma + (\psi + \psi_h)\gamma_h \cdots + (\psi + \cdots \psi_{h^{n-2}})\gamma_{h^{n-2}}].$$

Note that γ above does exist since, $\forall \gamma \in F_h - \{0\}$,

$$Tr_{F_h}^F(\gamma) = n\gamma \neq 0.$$

This concludes the proof of lemma 4.2.4. ∎

Lemma 4.2.6 (Multiplicative decomposition) *Assume now* α *and* β *are two automorphisms of* F, *satisfying* $\alpha^2 = \beta^2 = I_d$. *Let* $h = \alpha\beta$ *and* $f \in F$, *such that*

$$\begin{cases} N_{F_h}^F(f) = 1, \\ N_{F_\alpha}^F(f) = ff_\alpha = 1. \end{cases}$$

Then there exists $c \in F_\beta$, *satisfying*

$$f = \frac{c}{c_h}.$$

∎

Proof. From lemma 4.2.4, we know that there exists $a \in F$, with

$$f = \frac{a}{a_h}.$$

Hence

$$f_\alpha = \frac{a_\alpha}{a_\beta} = \frac{1}{f} = \frac{a_h}{a},$$

which yields in turn

$$\frac{a_\alpha}{a_h} = \frac{a_\beta}{a}. \tag{4.2.16}$$

Since

$$\frac{a_\alpha}{a_h} = \left(\frac{a}{a_\beta}\right)_\alpha,$$

equation (4.2.16) can be rewritten as

$$\left(\frac{a}{a_\beta}\right)_\beta = \left(\frac{a}{a_\beta}\right)_\alpha \quad \text{or} \quad \left(\frac{a}{a_\beta}\right)_h = \frac{a}{a_\beta}.$$

Introduce the sub-field

$$F_{\beta,\alpha} = F_\beta \cap F_\alpha.$$

We shall prove now that F_h is an algebraic extension of degree 2 of $F_{\beta,\alpha}$, generated by β (resp. α) with a group of automorphisms $(1, \beta)$ (resp. $(1, \alpha)$).

• First, if $z \in F_\beta \cap F_\alpha$, then $z = z_\beta = z_\alpha$, so that $z = z_h$, i.e. $F_h \supset F_\beta \cap F_\alpha$.

• Let now $z \in F_h$, i.e. $z = z_h$ or, equivalently, $z_\alpha = z_\beta$. Then $z + z_\beta \in F_\beta$ and $z + z_\beta = z + z_\alpha \in F_\alpha$, so that $z + z_\beta \in F_\beta \cap F_\alpha$.

• Similarly, $z z_\beta \in F_\beta \cap F_\alpha$.

One has thus shown that that z is a zero of an equation of order 2, with coefficients in $F_{\beta,\alpha}$. Moreover, fixing $x \in F_h$, $x \notin F_{\beta,\alpha}$, we obtain, for any $z \in F_h$,

$$z = u + vx,$$

where

$$u = \frac{x z_\alpha - x_\alpha z}{x - x_\alpha} \quad \text{and} \quad v = \frac{z - z_\alpha}{x - x_\alpha}.$$

It is easy to check that $u, v \in F_{\beta,\alpha}$. Consequently, F_h is an extention of order 2 of $F_{\beta,\alpha}$.

Since $N_{F_{\beta,\alpha}}^{F_h}\left(\frac{a_\beta}{a}\right) = 1$, we can apply lemma 4.2.4 to the function $\frac{a_\beta}{a}$ with $h = \alpha$. Hence, there exists $b \in F_h$ such that

$$\frac{b}{b_\beta} = \frac{a_\beta}{a},$$

which implies $ab \in F_\beta$. Putting $c = ab$, we get

$$\frac{c}{c_h} = \frac{ab}{a_h b_h} = \frac{a}{a_h} = f.$$

The proof of lemma 4.2.6 is concluded.   ∎

Remark 4.2.7   *It is also worth noting that, by construction, the element c found in lemma 4.2.6 has the form*

$$c = \sum_{i=0}^{n-1} \theta_{h^i} \prod_{j=0}^{i-1} f_{h^j},$$

with

$$\theta = \theta_\beta.$$

Simpler constructions exist for $n = 2, 3$.

- *When $n = 2$ and $f \neq -1$, we put*

$$c = 1 + f, \quad since \quad \frac{1+f}{1+f_h} = f.$$

- *For $n = 2$ and $f = -1$, we can take*

$$c = x - x_h.$$

- *For $n = 3$, we can choose c equal to one of the following elements*

$$\begin{cases} 1 + f + f_h, \\ x + x_h f + x_{h^2} f_h, \\ \dfrac{1}{x} + \dfrac{f}{x_h} + \dfrac{1}{x_{h^2}} f_h. \end{cases}$$

Lemma 4.2.8 (Additive decomposition) *Let α, β, h be as in lemma 4.2.6. Setting $\varepsilon = \pm 1$, let $u \in F$ such that*

$$\begin{cases} Tr^F_{F_h}(u) = 0, \\ u + \varepsilon u_\alpha = 0. \end{cases}$$

Then there exists $\gamma \in F$ satisfying $u = \gamma - \gamma_h$, with $\gamma = \varepsilon\gamma_\beta$. ∎

Proof.   It is sufficient to find a solution V of the equation

$$V - V_h = u,$$

which corresponds to an additive problem in lemma 4.2.4, and to put

$$\gamma = \frac{1}{2}(V + \varepsilon V_\beta).$$

Then clearly $\gamma = \varepsilon\gamma_\beta$, and

$$\gamma - \gamma_h = \frac{1}{2}(V - V_h + \varepsilon(V_\beta - V_\alpha)) = \frac{1}{2}(V - V_h + V - V_h) = u,$$

since the condition $u + \varepsilon u_\alpha = 0$ implies, instantiating $u = V - V_h$, that

$$V - V_h + \varepsilon(V_\alpha - V_\beta) = 0.$$

The case $\varepsilon = -1$ will be used only at the end of section 4.3. ∎

In the applications of lemmas 4.2.6 and 4.2.8, one will take

$$\alpha = \eta, \quad \beta = \xi, \quad h = \delta, \quad F = \mathbb{C}_Q(x, y),$$

whence $F_\xi = \mathbb{C}(x)$, $F_\eta = \mathbb{C}(y)$. Now we are in a position to formulate the final result of this section.

Theorem 4.2.9   *Let $N(f) = 1$. Then equation (4.2.5) has a rational solution if, and only if,*

$$\sum_{k=0}^{n-1} \psi_{\delta^k} \prod_{i=k+1}^{n-1} f_{\delta^i} = 0. \tag{4.2.17}$$

Moreover, under this condition, the solution is unique up to rational solutions of the system

$$u_\xi = u_\eta = u. \tag{4.2.18}$$

∎

Proof.   Coming back to the basic equation (4.2.5), we get from lemma 4.2.6

$$(c\pi)_\delta - c\pi = c_\delta \psi, \tag{4.2.19}$$

where $c \in \mathbb{C}(x)$ and $c_\delta \psi \in \mathbb{C}_Q(x, y)$. Setting $w = c\pi$, (4.2.5) finally reduces to

$$w_\delta - w = c_\delta \psi. \tag{4.2.20}$$

From the additive form of Hilbert's theorem, equation (4.2.20) has a solution in $\mathbb{C}_Q$ if, and only if, $Tr(c_\delta \psi) = 0$. With $c = fc_\delta$, we have

$$Tr(c_\delta \psi) = \sum_{k=0}^{n-1} (c_\delta \psi)_{\delta^k} = \sum_{k=0}^{n-1} \psi_{\delta^k} c_{\delta^{k+1}}$$

$$= c \sum_{k=0}^{n-1} \left(\frac{\psi_{\delta^k}}{\prod_{i=0}^{k} f_{\delta^i}} \right),$$

so that, using $N(f) = 1$, (4.2.17) is plainly equivalent to $Tr(c_\delta \psi) = 0$, which will be in force in the rest of the proof. Then lemma 4.2.8, with $\epsilon = 1$ ensures the existence of $\gamma \in \mathbb{C}(x)$, where

$$c_\delta \psi = \gamma - \gamma_\delta,$$

provided that $c_\delta \psi$ satisfies the second condition of lemma 4.2.8, which is tantamount to checking

$$c_\delta \psi + (c_\delta \psi)_\eta = 0.$$

But, from the definitions given in (4.2.6), we obtain precisely

$$c_\delta \psi + (c_\delta \psi)_\eta = c_\delta \left(\frac{r - r_\eta}{\varphi_\eta} + \frac{\varphi}{\varphi_\eta} \frac{r_\eta - r}{\varphi} \right) \equiv 0.$$

Now equation (4.2.20) becomes

$$w_\delta - w = \gamma - \gamma_\delta,$$

or

$$(w + \gamma)_\delta = w + \gamma.$$

Hence

$$w = -\gamma + u,$$

where u satisfies $u_\delta = u$. It follows that the rational solutions of (4.2.19) have the form

$$t = \frac{-\gamma + u}{c},$$

where u is itself a rational solution of the system of equations (4.2.18). The proof of theorem 4.2.9 is terminated. ∎

One of the main results of section 4.2 is that we could reduce the solution of the non homogeneous equation to that of a homogeneous one. This is summarized in the next theorem.

Theorem 4.2.10   *All rational solutions of the fundamental equations (4.2.1)-(4.2.3) can be classified as follows.*

(i) *If $N(f) \neq 1$, then there exists a unique rational solution ρ, which is given by (4.2.9).*

(ii) *If $N(f) = 1$, there are two cases:*

   – if condition (4.2.17) is not satisfied, then equation (4.2.5) has no rational solution;

   – if condition (4.2.17) holds, then any rational solution of (4.2.5) writes

$$\pi = t_0 + \frac{u}{c},$$

where c is rational, given by lemma 4.2.6 and remark 4.2.3, t_0 is the particular rational solution of (4.2.5) given by

$$t_0 = \frac{1}{2n} \sum_{i=1}^{n-1} (i - n) \left[\prod_{j=i}^{n-1} f_{\delta^j} \psi_{\delta^{i-1}} + \left(\prod_{j=i}^{n-1} f_{\delta^j} \psi_{\delta^{i-1}} \right)_\xi \right], \qquad (4.2.21)$$

and u is an arbitrary rational function of the composed function

$$\Delta^{(n)} \circ \Delta,$$

where Δ is the fractional linear transform defined by (3.3.4), and

$$\Delta^{(n)}(x) \overset{def}{=} \frac{\prod\limits_{j=0}^{j=n-1} \left(x - \wp\left(e + \frac{j\omega_2}{n}; \omega_1, \omega_2\right) \right)}{\prod\limits_{j=1}^{j=n-1} \left(x - \wp\left(\frac{j\omega_2}{n}; \omega_1, \omega_2\right) \right)}, \qquad (4.2.22)$$

denoting by

$$e = \frac{\alpha\omega_1}{2} + \frac{\beta\omega_2}{2n}$$

a zero of $\wp(\omega; \omega_1, \frac{\omega_2}{n})$, *with* $\alpha, \beta \in\ ]0, 1]$ *and either* α *or* β *is equal to 1.* ∎

Proof.   Assuming $N(f) = 1$ and condition 4.2.17, it is possible by lemma 4.2.8 to look for t_0 of the form

$$t_0 = -\frac{V + V_\xi}{2c},$$

where, instantiating $\gamma = 1$ in lemma 4.2.4, one chooses

$$V = \frac{1}{n}[(n-1)c_\delta\psi + (n-2)c_{\delta^2}\psi_\delta + \cdots + c_{\delta^{n-1}}\psi_{\delta^{n-2}}]$$

$$= \frac{1}{n}\sum_{i=1}^{n-1}(n-i)c_{\delta^i}\psi_{\delta^{i-1}}.$$

Iterating the relation $c = fc_\delta$, the above choice of V yields easily (4.2.21).

As for u, it follows by condition (4.2.18), since $u \in \mathbb{C}_Q$, that, on the universal covering $\mathbb{C}\omega$, $u(\omega)$ is an elliptic function with periods ω_1, ω_2, satisfying

$$\begin{cases} u = u_\xi \Rightarrow u(\omega) = u(-\omega) \\ u = u_\delta \Rightarrow u(\omega) = u(\omega + \omega_3). \end{cases}$$

From chapter 3 and section 4.1, we know that $\omega_3 = \frac{k}{n}\omega_2$, where k and n are relatively prime numbers. There exists an integer k', such that

$$kk' = 1 \quad \mod n,$$

and hence u is elliptic with periods ω_1 and $\frac{\omega_2}{n}$. Since u is even, we know (see e.g. [5]) that

$$u(\omega) = \Delta_0 \circ \wp\left(\omega; \omega_1, \frac{\omega_2}{n}\right),$$

where Δ_0 is an unspecified rational function. Also, using some by the basic properties of the Weierstrass $\wp$-function (see [5]), we can write

$$\wp\left(\omega; \omega_1, \frac{\omega_2}{n}\right) = \Delta^{(n)} \circ \wp(\omega; \omega_1, \omega_2),$$

where $\Delta^{(n)}$ is given by (4.2.22). Since from the uniformisation (3.3.4) $\wp$ is a fractional linear transform of x, the proof of the theorem is concluded.   ∎

Corollary 4.2.11 *The group* $\mathcal{H}$ *is of finite order* n *if, and only if,*

$$\mathbb{C}(x) \cap \mathbb{C}(y) \neq \mathbb{C}.$$

∎

Proof.   Indeed if n is finite, the set $\mathbb{C}(x) \cap \mathbb{C}(y) \neq \mathbb{C}$ does coincide with the set of rational solutions of (4.2.18). On the other hand, if $n = \infty$, then $\omega_1, \omega_2, \omega_3$ are linearly independant on $\mathbb{Z}$ and (4.2.18 admits only constant solutions.   ∎

4.3 Algebraic Solutions

The general and standard notion of algebraic function was recalled in section 2.1.2. Consider the equation

$$t_\delta - ft = \psi, \tag{4.3.1}$$

where t is subject to the constraint

$$t = \varepsilon t_\xi, \quad \varepsilon = \pm 1. \tag{4.3.2}$$

Since by (4.2.6) $ff_\eta = 1$, and by (4.3.2) $t_\delta = \varepsilon t_\eta$, a necessary condition for (4.3.1) to be solvable is

$$\psi + \varepsilon \psi_\eta f = 0. \tag{4.3.3}$$

In this section one will assume (4.3.3). The reason for condition (4.3.2) with $\varepsilon = -1$ will appear at the end of subsection 4.3.2. As before, one must separate the cases

$$N(f) = 1 \quad \text{and} \quad N(f) \neq 1.$$

4.3.1 The Case $N(f) = 1$

Recall that by lemma 4.2.6 there exists $c \in \mathbb{C}(x)$, such that $f = \dfrac{c}{c_\delta}$.

Theorem 4.3.1   *Equations (4.3.1) and (4.3.2) have a non-zero algebraic solution if, and only if,*

$$\sum_{k=0}^{n-1} \psi_{\delta^k} \prod_{i=k+1}^{n-1} f_{\delta^i} = 0,$$

which is nothing else but condition (4.2.17). Indeed, when (4.2.17) holds, all solutions are algebraic and have the form

$$t = t_1 + \frac{w}{c},$$

where $t_1 \in \mathbb{C}_Q$ is given by

$$t_1 = \frac{1}{2n} \sum_{i=1}^{n-1} (i-n) \left[\prod_{j=i}^{n-1} f_{\delta^j} \psi_{\delta^{i-1}} + \varepsilon \left(\prod_{j=i}^{n-1} f_{\delta^j} \psi_{\delta^{i-1}} \right)_\xi \right], \tag{4.3.4}$$

and w is an algebraic solution of the system

$$w_\delta = \varepsilon w_\xi = w.$$

■

Proof.   Setting $w = ct$ and substituting $c = fc_\delta$, equations (4.3.1) and (4.3.2) rewrite

$$w_\delta - w = c_\delta \psi.$$

Define

$$\begin{cases} \psi_2 & = \dfrac{1}{n} Tr(c_\delta \psi), \\ \psi_1 & = c_\delta \psi - \dfrac{1}{n} Tr(c_\delta \psi), \end{cases}$$

Then all solutions of (4.3.1), (4.3.2) have the form

$$w = w_1 + w_2 + U, \tag{4.3.5}$$

where

- The function $w_i, i = 1, 2$, is a particular solution of the system

$$\begin{cases} w_\delta - w = \psi_i, & i = 1, 2, \\ w_\xi = \varepsilon w. \end{cases}$$

  In addition, from lemma 4.2.8 and lemma 4.3.2 to be proved below, w_1 exists in $\mathbb{C}_Q$ since

$$Tr(\psi_1) = 0, \quad \psi_1 + \varepsilon(\psi_1)_\eta = 0.$$

  The computation of w_2 will be carried out after the proof of lemma 4.3.5.

- The function U satisfies the homogeneous equation

$$U_\delta = \varepsilon U_\xi = U.$$

  In fact, U is algebraic and completely characterized by means of lemma 4.3.3.

Lemma 4.3.2   *We have*

$$\psi_i + \varepsilon(\psi_i)_\eta = 0, \quad i = 1, 2.$$

■

Proof.   There exists g such that

$$c_\delta \psi = g_\eta - \varepsilon g.$$

Indeed, using $c = fc_\delta = c_\xi$, we get by (4.3.3) $\varepsilon(c_\delta \psi)_\eta + c_\delta \psi = 0$, and one can simply choose

$$g = \frac{c_\delta \psi - \varepsilon(c_\delta \psi)_\eta}{4} = \frac{-\varepsilon c_\delta \psi}{2}.$$

Then

$$\psi_2 = \frac{1}{n}Tr(g_\eta) - \varepsilon\frac{1}{n}Tr(g).$$

Since, from an earlier derivation, $Tr(g_\eta) = (Tr(g))_\eta$, one can write

$$\psi_2 = \frac{1}{n}[Tr(g)_\eta - \varepsilon Tr(g)]$$

and

$$\psi_2 + \varepsilon(\psi_2)_\eta = 0.$$

The proof of the lemma is concluded, just remembering that, by definition,

$$\psi_1 = c_\delta\psi - \psi_2.$$

■

Lemma 4.3.3 *On the universal covering $\mathbb{C}_\omega$, let us consider the field of meromorphic functions with period ω_1. Then, in this field, any solution of the equation*

$$w = w_\delta \tag{4.3.6}$$

is an algebraic function of x. ■

Proof.   On $\mathbb{C}_\omega$, (4.3.6) becomes

$$w(\omega) = w(\omega + \omega_3).$$

Thus w is a function with periods ω_1, ω_3, so that

$$w(\omega) = \Delta_0\big(\wp(\omega;\omega_1,\omega_3)\big) + \wp'(\omega;\omega_1,\omega_3)\Delta_1\big(\wp(\omega;\omega_1,\omega_3)\big),$$

where Δ_0, Δ_1 are rational functions. Since $n\omega_3 = k\omega_2$, using the functions $\Delta^{(n)}$ introduced in the proof of theorem 4.2.10, we have

$$\begin{cases} \wp(\omega;\omega_1,\omega_3) = \Delta^{(n)} \circ \wp(\omega;\omega_1,k\omega_2), \\ \wp(\omega;\omega_1,\omega_2) = \Delta^{(k)} \circ \wp(\omega;\omega_1,k\omega_2). \end{cases}$$

Since $\wp'$ is given by relation (3.3.1) (with the convenient arguments), and x is a homographic function of $\wp$, lemma 4.3.3 is proved. ■

Corollary 4.3.4 *There is en effective construction of the algebraic solutions of (4.2.18)* ■

Proof.   Let

$$
\begin{cases}
u^*(\omega) = u\left(\omega + \dfrac{\omega_2}{2}\right), \\
x^*(\omega) = x\left(\omega + \dfrac{\omega_2}{2}\right).
\end{cases}
$$

Then, by (4.2.18), u^* is even, elliptic with periods ω_1, ω_3 and there exists Δ_0 rational, such that

$$
u^*(\omega) = \Delta_0(\wp(\omega; \omega_1, \omega_3)).
$$

Consequently

$$
u(\omega) = \Delta_0 \circ \wp\left(\omega - \frac{\omega_2}{2}; \omega_1, \omega_3\right).
$$

With $s \overset{\text{def}}{=} \wp\left(\omega - \dfrac{\omega_2}{2}; \omega_1, k\omega_2\right)$, and using the functions $\Delta^{(k)}, \Delta^{(n)}$ introduced lemma 4.3.3, we have

$$
\begin{cases}
u(\omega) = \Delta_0 \circ \Delta^{(n)}(s), \\
x^*(\omega) = \Delta_2 \circ \Delta^{(k)}(s),
\end{cases}
$$

where Δ_2 is the fractional linear transform given by (3.3.3). Then (see e.g [5]) there exists a fractional linear transform Δ_3, such that

$$
\wp(\omega; \omega_1, \omega_2) = \Delta_3 \circ \wp\left(\omega - \frac{\omega_2}{2}; \omega_1, \omega_2\right).
$$

Finally

$$
\begin{cases}
u(\omega) = \Delta_0 \circ \Delta^{(n)}(s), \\
x(\omega) = \Delta_2 \circ \Delta_3 \circ \Delta^{(k)}(s).
\end{cases}
$$

It is interesting to remark that, when $k = 1$, any solution u of (4.2.17) are in fact rational functions of x, since $\Delta^{(1)}$ is the identity. ∎

Continuing with the proof of theorem 4.3.1, note that, whenever $\psi_2 = 0$ (which corresponds to the condition of the theorem), it suffices to take $w_2 \equiv 0$.
It remains to prove that, for $\psi_2 \neq 0$, (4.3.1) and (4.3.2) have no algebraic solution. To this end, let us first find the function w_2 in considering the equations

$$
w_\delta - w = \psi_2, \qquad w_\xi = \varepsilon w,
$$

on the universal covering $\mathbb{C}_\omega$. We shall show that

$$
w_2(\omega) \overset{\text{def}}{=} \psi_2(\omega)\widetilde{\Phi}(\omega)
$$

is non-algebraic, where

$$
\Phi(\omega) \overset{\text{def}}{=} \frac{\omega_1}{2i\pi}\zeta(\omega; \omega_1, \omega_3) - \frac{\omega}{i\pi}\zeta\left(\frac{\omega_1}{2}; \omega_1, \omega_3\right) \overset{\text{def}}{=} \widetilde{\Phi}\left(\omega + \frac{\omega_2}{2}\right), \tag{4.3.7}
$$

and $\zeta(\omega)$ denotes the classical Weierstrass ζ-function (see [38]). An intermediate lemma is needed.

Lemma 4.3.5   *The function $\Phi(w)$ given by (4.3.7) is meromorphic in $\mathbb{C}_\omega$, odd, periodic with period ω_1, and satisfies*

$$\Phi(\omega + \omega_3) = \Phi(\omega) + 1.$$

■

Proof. Φ is meromorphic and odd since ζ has both these properties. To prove that Φ has period ω_1 and satisfies the equation of the lemma, it suffices to use (see e.g. [5])

$$\zeta(\omega + \omega_i) = \zeta(\omega) + 2\zeta\left(\frac{\omega_i}{2}\right), \quad i \in \{1, 3\},$$

together with Legendre's identity

$$\omega_1 \zeta\left(\frac{\omega_3}{2}\right) - \omega_3 \zeta\left(\frac{\omega_1}{2}\right) = \pi i.$$

Lemma 4.3.5 is proved.   ■

From the latter two lemmas, it follows that the product $\psi_2 \widetilde{\Phi}$ is meromorphic, periodic with period ω_1, and

$$
\begin{cases}
(w_2)_\delta = \psi_2(\omega)\widetilde{\Phi}(\omega + \omega_3) = w_2(\omega) + \psi_2(\omega), \\
(w_2)_\xi = (\psi_2(\omega))_\xi \Phi\left(\frac{\omega_2}{2} - \omega\right) = -(\psi_2(\omega))_\xi \widetilde{\Phi}(\omega) = \varepsilon w_2(\omega),
\end{cases}
$$

where we have used $(\psi_2)_\xi = (\psi_2)_\eta$ and lemma 4.3.2. Finally, w_2 is a solution of the equation

$$w_\delta - w = \psi_2,$$

subject to the constraint $w_2 = \varepsilon(w_2)_\xi$. On the other hand, in (4.3.5), w_1 and U are algebraic, so that w is algebraic if, and only if, w_2 is algebraic. Consequently, as ψ_2 is rational, it remains to show that $\Phi(\omega - \omega_2/2)$ is not algebraic in $x(\omega)$. To that end, we exploit the fact that Φ has a linear growth in the ω_2-direction. Recalling that $n\omega_3 = k\omega_2$, define, for all $m \in \mathbb{Z}$, the quantities

$$\alpha_m = mk\omega_2 = mn\omega_3.$$

Then, by lemma 4.3.5, we get, since $x(\omega)$ is elliptic with periods ω_1, ω_2,

$$
\begin{cases}
x(\omega) = x(\omega + \alpha_m), \\
\Phi(\omega + \alpha_m) = \Phi(\omega) + mn, \quad \forall m \in Z_+.
\end{cases}
$$

Thus, for a fixed x, Φ takes an infinite number of values, and therefore cannot be algebraic in x. The proof of theorem 4.3.1 is terminated.   ■

4.3.2 The Case $N(f) \neq 1$

Theorem 4.3.6   *Equations (4.2.2), (4.2.5) have an algebraic non-rational solution if, and only if,*

$$N(f) = -1.$$

The solution has the form

$$\pi = \rho + \frac{u}{c},$$

where ρ is given by (4.2.9), u is an algebraic solution of equation (4.3.6), and c is algebraic given by

$$c = \sum_{i=0}^{n-1} (\theta_{\delta i} - \theta_{\delta i + n}) \prod_{j=0}^{i=1} f_{\delta i}, \qquad (4.3.8)$$

where $\theta \in F_\xi$ is chosen to ensure $c \neq 0$ (in particular, $\theta \notin \mathbb{C}_Q(x,y)$), F being an algebraic extension of $\mathbb{C}_Q(x,y)$ to be precisely stated in the course of the proof. ∎

Proof.   Since $N(f) \neq 1$, then (4.2.2), (4.2.5) have a rational solution. Hence, we can consider only the homogeneous equation of the type (4.2.15). Let us suppose that the system

$$\begin{cases} t_\delta &= ft, \\ t_\xi &= t, \end{cases} \qquad (4.3.9)$$

where f satisfies

$$f f_\eta = 1,$$

have an algebraic solution, i.e. $P(t, x) \equiv 0$ for some polynomial P. The purpose is to find, on the universal covering $\mathbb{C}_\omega$, meromorphic solutions of (4.3.9), with period ω_1.

Notation For any function $g : z \mapsto g(z)$, one will write $g'_z \overset{\text{def}}{=} \dfrac{dg}{dz}$ for the derivative of g with respect to z, or simply g' if there is no ambiguity.

By derivation of $P(t, x) = 0$ with respect to x, we see that t'_x is rational of t and x, and hence the ratio $\dfrac{t'_x}{t}$ is algebraic in x. Moreover, since $x(\omega)$ is an elliptic function of ω, x'_ω is also elliptic with the same periods, and is therefore algebraic in x. Using then the elementary equality

$$\frac{t'_\omega}{t} = \frac{x'_\omega t'_x}{t},$$

we deduce also that $\dfrac{t'_\omega}{t}$ is algebraic in x. Introducing the following logarithmic derivatives

$$u = \frac{t'_\omega}{t} \quad \text{and} \quad v = \frac{f'_\omega}{f},$$

we get from (4.3.9) that u is an algebraic solution of

$$\begin{cases} u_\delta - u = v, \\ u_\xi + u = 0, \\ v_\eta - v = 0. \end{cases} \tag{4.3.10}$$

Putting in (4.3.1), (4.3.2), $f = 1$, $\psi = v$, $\varepsilon = -1$, one can apply theorem 4.3.1, just replacing in its statement f by 1 and ψ by v. This yields

$$Tr(v) = 0.$$

But

$$Tr(v) = Tr\left(\frac{f'}{f}\right) = \frac{N(f)'}{N(f)},$$

since

$$\delta^k(\omega) = \omega + k\omega_3,$$

which in turn implies

$$\frac{d\delta^k(\omega)}{d\omega} = 1.$$

Then $Tr(v) = 0$ implies $N(f) = K$, where K is some constant and we have the following chain of equalities

$$f f_\eta = 1 \Rightarrow N(f)N(f_\eta) = 1 \Rightarrow N(f)(N(f))_\eta = 1 \Rightarrow K^2 = 1 \Rightarrow K = -1,$$

since $K \neq 1$ and this proves the *if* assertion of the theorem.

Now, let us consider the homogeneous equation, with $N(f) = -1$. We intend to prove that this equation has an algebraic solution. For this, let us note that

$$\prod_{i=0}^{2n-1} f_{\delta^i} = \left(\prod_{i=0}^{n-1} f_{\delta^i}\right)^2 = (N(f))^2 = 1,$$

as $f \in \mathbb{C}_Q(x, y)$ and $\delta^n = I_d$.

We would like to get a kind of Hilbert's factorization of f. To this end, consider the field F of elliptic functions with periods ω_1 and $2n\omega_3$. Since $n\omega_3 = k\omega_2$, $\mathbb{C}_Q(x, y)$ can be considered as a subfield of this field. For any $u \in F$, δ is then defined by

$$u_\delta(\omega) = u(\omega + \omega_3).$$

The cyclic group generated by δ on F is finite of order $2n$. Clearly, $f \in F$ and moreover

$$N_{F_\delta}^F(f) = 1,$$

where F_δ is the subfield of elliptic functions with periods ω_1 and ω_3. By lemma 4.2.6, since

$$N_{F_\eta}^F(f) = f f_\eta = 1,$$

there exists $c \in F_\xi$, where F_ξ is the subfield of elements of F invariant with respect to ξ, such that

$$f = \frac{c}{c_\delta}.$$

Then system (4.3.9) becomes

$$\begin{cases} (tc)_\delta &= tc, \\ (tc)_\xi &= tc, \end{cases}$$

so that tc is elliptic with periods ω_1, ω_3, and, applying lemma 4.3.3 with $\varepsilon = 1$, algebraic with respect to x.

The last point is to show that c is also algebraic . But this follows directly from the two following properties:

- c is an elliptic function with periods ω_1 and $2n\omega_3 = 2k\omega_2$ which can be obtained as in remark 4.2.7 and has the form (4.3.8).

- x is an elliptic function with primitive periods ω_1 and ω_2.

The proof of theorem 4.3.6 is concluded.   ■

4.4 Final Form of the General Solution

In section 4.3, we proved *en passant* the following theorem which gives the general solution when $N(f) = 1$.

Theorem 4.4.1   *If $N(f) = 1$, then the general solution of the fundamental equations (4.2.2), (4.2.5) has the form*

$$\pi = w_1 + w_2 + \frac{w}{c}, \tag{4.4.1}$$

where

– the function c is defined by lemma 4.2.6;

– the function w_1 is given by formula (4.2.21) in which ψ is replaced by

$$\psi - \frac{1}{n} \sum_{k=0}^{n-1} \left(\frac{\psi_{\delta^k}}{\prod_{i=1}^k f_{\delta^i}} \right);$$

– the function w_2 is given by

$$w_2(\omega) = \frac{1}{n} \sum_{k=0}^{n-1} \left(\frac{\psi_{\delta^k}}{\prod_{i=1}^{k} f_{\delta^i}} \right)(\omega) \, \widetilde{\Phi}(\omega). \tag{4.4.2}$$

– *the function w is an algebraic function satisfying $w = w_\xi = w_\delta$.* ∎

Now we shall deal with the case $N(f) \neq 1$, proceeding along similar patterns. First let us find the solutions of the homogeneous equation

$$t_\delta = ft, \quad \text{with} \quad t(\omega) = t(-\omega + \omega_2), \quad \forall \omega \in \mathbb{C}_\omega. \tag{4.4.3}$$

For convenience, we shall put

$$\frac{t'_\omega}{t} = u. \tag{4.4.4}$$

Then u satisfies (take the logarithm in (4.4.4) and differentiate)

$$u_\delta - u = \frac{f'_\omega}{f} \equiv \chi, \quad u_\xi = -u, \tag{4.4.5}$$

with

$$\chi = \chi_\eta. \tag{4.4.6}$$

Remembering that

$$\eta(\omega) = \omega_2 + \omega_3 - \omega,$$

equation (4.4.6) is a consequence of the following the chain of equivalences (see lemma 4.2.8 with $\varepsilon = -1$):

$$f f_\eta = 1 \Leftrightarrow f(\omega) f(\omega_2 + \omega_3 - \omega) = 1 \Leftrightarrow \chi(\omega) = \chi(\omega_2 + \omega_3 - \omega) \Leftrightarrow \chi - \chi_\eta = 0.$$

We put now, as in section 4.2,

$$u = \frac{1}{n} Tr(\chi) \widetilde{\Phi} + a \tag{4.4.7}$$

where $\widetilde{\Phi}$ is given in (4.3.7) and a is such that

$$\begin{cases} a_\delta - a = \chi - \frac{1}{n} Tr(\chi) \overset{\text{def}}{=} \widetilde{\chi}, \\ a_\xi = -a. \end{cases} \tag{4.4.8}$$

The function a exists, just putting into lemma 4.2.8, with $\varepsilon = -1$,

$$u = \widetilde{\chi} \quad \text{and} \quad a = -\gamma.$$

Then, as before, we obtain

$$u_\xi = -u. \tag{4.4.9}$$

The solution t of equation (4.4.4) is meromorphic if, and only if, all residues of the meromorphic function u are integers. This can be checked by choosing

$$a = \frac{1}{n} \sum_{k=0}^{n-1} k \widetilde{\chi}_{\delta^k}.$$

On the other hand,

$$\widetilde{\chi} = \frac{f'}{f} - \frac{1}{n} Tr \left(\frac{f'}{f} \right),$$

so that a can be rewritten as

$$a = \frac{1}{n} \sum_{k=0}^{n-1} \left(k - \frac{n-1}{2} \right) \left(\frac{f'}{f} \right)_{\delta^k}.$$

The general solution of (4.4.5) is of the form

$$u = \frac{1}{n} Tr(\chi) \widetilde{\Phi} + a - R, \tag{4.4.10}$$

where R is a solution of the homogeneous system

$$\begin{cases} R_\delta = R & \Longleftrightarrow R(\omega + \omega_3) = R(\omega), \\ R_\xi = -R & \Longleftrightarrow R(-\omega + \omega_2) = -R(\omega). \end{cases}$$

One will choose R so that the residues become integer-valued.

Lemma 4.4.2   *There exists an elliptic function R, with periods ω_1, ω_3, such that*

$$R(-\omega + \omega_2) = -R(\omega),$$

and all residues of u in (4.4.10) are integer-valued.   ■

Proof.   Let us consider the fundamental rectangle Π with vertices b, $b + \omega_1$, $b + \omega_3$, $b + \omega_1 + \omega_3$, where b will be fixed later on. We shall proceed in steps.

• First, one will show that the difference of residues at the points z and $z + \omega_3$, for any z, is an integer. Let A be a small circle around z. Then

$$\int_{\delta A} u \, d\omega - \int_A u = \int_A \frac{1}{n} (Tr(\chi))(\widetilde{\Phi} + 1) d\omega - \int_A \frac{1}{n} (Tr(\chi)) \widetilde{\Phi} d\omega + \int_A (a_\delta - a) d\omega$$

$$= \int_A \frac{1}{n} (Tr(\chi)) d\omega + \int_A \widetilde{\chi} d\omega = \int_A \chi d\omega = \int_A \frac{f'}{f} d\omega = 2iK\pi,$$

where $|K|$ is the multiplicity of the zero or of the pole.

• Secondly, one must prove the sum of residues of (4.4.7) inside Π is an integer. Take now

$$b = -\frac{\omega_3}{2} - \frac{\omega_2}{2}.$$

Then, setting $L = \left[\frac{\omega_3}{2} - \frac{\omega_1}{2}, \frac{\omega_3}{2} + \frac{\omega_1}{2}\right]$, we have (since the sum of the integrals along imaginary directions cancel)

$$\int_{\Pi} u d\omega = \int_{L} u d\omega - \int_{\delta^{-1}L} d\omega = \int_{\delta^{-1}L} \chi d\omega = 2\pi i \, [\arg(f)]_{\delta^{-1}L} = 2\pi i \, [\arg(f_{\delta^{-1}})]_{L},$$

where $[\arg(z)]_L$ denotes the variation of the argument of z in traversing the contour L in the positive direction. The image of L (which is invariant w.r.t. η) under the projection $x(\lambda(\omega))$ onto $\mathbb{C}_x$ is a closed curve, coinciding with the cut $[y_1, y_2]$. Thus the right hand-side above is an integer.

• Thirdly, it is necessary to show that R exits with $R(-\omega + \omega_2) = -R(\omega)$. But this is clear from (4.4.9) and proof of lemma 4.4.2 is complete. ∎

In fact we want to give a somehow more explicit construction of R, noting that u in (4.4.7) is periodic with periods ω_1, $k\omega_2$, provided that $\omega_3 = \frac{k}{n}\omega_2$. To this end, let us introduce two families a_i, b_i, $1 \le i \le r$, which are respectively the poles and the zeros of f in the rectangle $\tilde{\Pi} =]0, \omega_1[\times]0, \omega_2[$, and consider the points

$$a_i + \ell\omega_2 \mod \omega_3, \quad \text{and} \quad b_i + \ell\omega_2 \mod \omega_3, \quad \forall 1 \le i \le r, 1 \le \ell \le k - 1.$$

We choose these points to be poles and zeros of f respectively, and having in addition their real component on the segment $]0, \frac{\omega_3}{2}[$. Then the residues of R at an arbitrary point α are given by the formula

$$\text{Res}_{\alpha} R = K_{\alpha} + \frac{\kappa}{n}\left(\tilde{\Phi}(\alpha) - \frac{n-1}{2}\right),$$

where

$$\kappa = \begin{cases} -1 & \text{if } \alpha \text{ is a pole of } f, \\ 1 & \text{if } \alpha \text{ is a zero of } f, \end{cases}$$

and

$$\begin{cases} \text{Res}_0\tilde{\Phi} = \dfrac{\omega_1}{2i\pi}, \\ \text{Res}_0 R = \dfrac{\text{Res}_0\tilde{\Phi}}{n}\left[\dfrac{(Nf)'}{Nf}\right]_{|\omega=0}, \\ \tilde{R}(\omega) \stackrel{\text{def}}{=} R\left(\omega - \dfrac{\omega_2}{2}\right). \end{cases}$$

Above, α takes one of the values

$$a_i + \ell\omega_2, \ b_i + \ell\omega_2, \ i = 1, \ldots, r \, ; \ \ell = 0, \ldots, k - 1,$$

and the integers K_{α} are chosen to have

$$\begin{cases} 2(\sum K_{ij} + \sum \widetilde{K}_{ij}) &= [\arg(f_{\delta-1})], \\ \sum K_{ij} + \sum \widetilde{K}_{ij} &= -[\arg(\varphi_\xi)], \end{cases}$$

where

$$\begin{cases} K_{ij} = K_\alpha, & \text{for} \quad \alpha = a_i + j\omega_2, \\ \widetilde{K}_{ij} = K_\alpha, & \text{for} \quad \alpha = b_i + j\omega_2, \end{cases}$$

and the variation of the argument is taken along the cut $]y_1, y_2[$ in the direction corresponding to the direction on L. Note that one could get rid of the K_α's, i.e. one could put $K_\alpha \equiv 0$ by introducing at the very beginning, in place of f and φ, the following functions:

$$\hat{\varphi}(x) = \varphi(x)x^{-\arg[\varphi_\xi]}, \quad \hat{f} = \frac{\hat{\varphi}}{\hat{\varphi}_\eta}.$$

Then it is convenient to write $\widetilde{R}$ in the integral form

$$\widetilde{R}(\omega) = \frac{1}{2\pi i} \left[\int_\Pi \frac{\wp'(t)\hat{u}(t)dt}{\wp(t) - \wp(\omega)} + \omega_1\zeta(\omega) \right]$$

where $\wp, \zeta$ are the well known Weierstrass functions with periods ω_1, ω_3. This yields the more explicit formula

$$\begin{aligned} \widetilde{R}(\omega) = \quad & \sum \frac{\wp'(\omega)}{\wp(\omega) - \wp(b_i + j\omega_2)} \left(K_{ij} + \frac{1}{n}(\widetilde{\Phi}(b_i + j\omega_2) - \frac{n-1}{2}) \right) \\ - \quad & \sum \frac{\wp'(\omega)}{\wp(\omega) - \wp(a_i + j\omega_2)} \left(\widetilde{K}_{ij} + \frac{1}{n}(\widetilde{\Phi}(a_i + j\omega_2) - \frac{n-1}{2}) \right) \\ - \quad & \frac{1}{n} \left[\frac{(Nf)'}{Nf} \right]_{|\omega=0} \frac{\omega_1 \, \wp'(\omega)}{2\pi i \, \wp(\omega)}. \end{aligned}$$

Hence we come to the following result, similar to the one previously obtained in theorem 4.4.1.

Theorem 4.4.3   *If $N(f) \neq 1$ then the general solution of the fundamental equations (4.3.1), (4.3.2) has the form*

$$\pi = \rho + \theta \exp \left(\int \left(\frac{1}{n} Tr(\chi)\widetilde{\Phi} - \widetilde{R} \right) d\omega \right),$$

where $\rho \in \mathbb{C}(x)$ was found in section 4.2, χ, $\widetilde{\Phi}$, $\widetilde{R}$ are defined above and θ is an algebraic function satisfying

$$\theta = \theta_\xi = \theta_\eta.$$

■

4.5 The Problem of the Poles and Examples

In sections 4.2 - 4.4, we have completely described the structure of the solutions of the fundamental equation in the case of a finite group. However the problem of specifying a unique probabilistic solution was left opened. In other words, we must choose solutions π, $\widetilde{\pi}$ without poles in the unit circle.

Consider first the case $N(f) = 1$ (see theorem 4.4.1), and assume for a while the functions c, w_1, w_2 have already been computed. Then we must choose U, so that both functions

$$\pi = w_1 + w_2 + \frac{U}{c}, \quad \widetilde{\pi} = -\frac{q\pi + q_0\pi_{00}}{\widetilde{q}},$$

have no poles in the unit circle. We know that this is possible if, and only if, the system is ergodic (see chapter 1). In our opinion, the problem of explicit finding the poles of U is of a computational nature. It is not excluded there might be deeper insights into the problem of poles, as can be guessed from the examples presentedbelow.

Similarly, when $N(f) \neq 1$, one must find the function θ introduced in theorem theorem 4.4.3.

4.5.1 Rational Solutions

Here one will analyze miscellaneous random walks, which have been encountered in several studies related to computer models.

4.5.1.1 Reversible Random Walks Take an ergodic random walk, satisfying the following reversibility conditions [40]

$$\pi_\alpha p_{\alpha\beta} = \pi_\beta p_{\beta\alpha}, \ \forall \alpha, \beta \in \mathbb{Z}_+^2. \tag{4.5.1}$$

It is easy to prove, using (4.5.1), that the generating functions are rational. For instance, if $p'_{-1,0} \neq 0$ then

$$\pi_{i0} = C \left(\frac{p'_{1,0}}{p'_{-1,0}} \right)^i.$$

4.5.1.2 Simple Examples of Nonreversible Random Walks Suppose that q and $\widetilde{q}$ can be expressed as

$$\begin{cases} q(x, y) = xa(x, y)a_1(x)a_2(y) & \text{mod } Q(x, y), \\ \widetilde{q}(x, y) = ya(x, y)\widetilde{a}_1(x)\widetilde{a}_2(y) & \text{mod } Q(x, y), \end{cases} \tag{4.5.2}$$

for some rational functions a, a_i, $\widetilde{a}_i$. Then, putting

$$\pi_1(x) = \frac{x\pi(x)a_1(x)}{\widetilde{a}_1(x)}, \quad \widetilde{\pi}_1(y) = \frac{y\widetilde{\pi}(y)a_2(y)}{\widetilde{a}_2(y)},$$

the fundamental equation takes the form

$$\pi_1(x) + \widetilde{\pi}_1(y) + P(x,y) = 0, \tag{4.5.3}$$

where

$$P(x,y) = \pi_{00}q_0(x,y)\frac{\widetilde{a}_1(x)a_2(y)}{a(x,y)}.$$

Eliminating $\widetilde{\pi}_1$ in (4.5.3), we obtain

$$(\pi_1)_\delta - \pi_1 = P - P_\eta.$$

Let us suppose hereafter the group is of order 4. Since $f = 1 = N(f)$, the latter equation admits one rational solution if, and only if,

$$0 = Tr(P - P_\eta) = P - P_\eta + P_\delta - P_\xi,$$

i.e. $P + P_\delta \in \mathbb{C}(x) \cap \mathbb{C}(y)$. When the latter condition holds, it also occurs frequently that P has the additive representation

$$P(x,y) = s(x) + \widetilde{s}(y),$$

in which case (4.5.3) yields

$$\begin{cases} \pi_1(x) = -s(x) + u(x), \\ \widetilde{\pi}_1(y) = -\widetilde{s}(y) - \widetilde{u}(y), \end{cases}$$

where u satisfies (4.2.18), i.e. $u \in \mathbb{C}(x) \cap \mathbb{C}(y)$, so that $u(x) = \widetilde{u}(y) \mod Q(x,y)$. Then

$$\begin{cases} \pi(x) = \left[-s(x) + u(x)\right]\dfrac{\widetilde{a}_1(x)}{xa_1(x)}, \\ \widetilde{\pi}(y) = -\left[\widetilde{s}(y) + \widetilde{u}(y)\right]\dfrac{\widetilde{a}_2(x)}{ya_2(y)}. \end{cases}$$

When they exist, the functions $u, \widetilde{u}$ are unique and, in this case, π and $\widetilde{\pi}$ must have no pole inside the unit disk.

It is not difficult to construct random walks fitting (4.5.2), with jumps from the boundaries not exceeding 2, and which are not reversible but have a rational $\pi(x)$. For instance, consider the random walk shown on figure 4.5.1, where

$$\begin{cases} Q(x,y) &= p_{-1,1}y^2 + (p_{10}x^2 - x + p_{-1,0})y + p_{-1,-1} \\ q(x,y) &= xq_0(x,y) = x(y^2p_{02}^0 + yp_{01}^0 - 1), \\ \widetilde{q}(x,y) &= y(x-1). \end{cases}$$

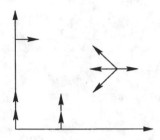

Fig. 4.5.1.

and we have

$$\frac{x\pi(x)}{x-1} + \frac{y\widetilde{\pi}(y)}{r(y)} + \frac{\pi_{00}}{x-1} = 0,$$

so that, setting

$$\pi_1(x) = \frac{x\pi(x) + \pi_{00}}{x-1}, \quad \pi_2(y) = \frac{y\widetilde{\pi}(y)}{r(y)},$$

we get

$$\pi_1(x) + \pi_2(y) = 0. \tag{4.5.4}$$

In (4.5.4), we have clearly $N(f) = 1$ and the trace is identically 0. Since one checks the group is of order 4, i.e $\omega_3 = \frac{\omega_2}{2}$, the remark made at the end of the proof of corollary 4.3.4 implies directly that

$$\frac{x\pi(x) + \pi_{00}}{x-1} \in \mathbb{C}(x) \cap \mathbb{C}(y).$$

In addition, the expression of Q shows that $\mathbb{C}(x) \cap \mathbb{C}(y)$ is generated by the element $yp_{-1,1} + \frac{p_{-1,-1}}{y}$, or equivalently by $x(1 - p_{10}x)$. Hence

$$\frac{x\pi(x) + \pi_{00}}{x-1} = \frac{y\widetilde{\pi}(y)}{(1-y)(yp_{02}^0 + 1)} = R\big(x(1 - p_{10}x)\big),$$

where R denotes a rational function, which will be exactly determined. To this end, we must refer to some properties concerning the branch $Y_0(x)$ and the analytic continuation of the functions $\pi, \widetilde{\pi}$ (see sections 2.3.1, 5.3 and also [26]). It emerges in particular that the curve $\mathcal{L}$ and $\mathcal{M}$ (introduced in section 5.3) are respectively a circle and a straight line, and also that

$$\frac{x\pi(x) + \pi_{00}}{x-1}$$

cannot have more than one pole, which shows that R is in fact a fractional linear transform of the form

$$R(t) = \frac{at + b}{t + p_{10} - 1}.$$

Straightforward calculations yield

$$\begin{cases} \pi(x) = \dfrac{\pi_{00} p_{10}}{1 - p_{10} - p_{10}x}, \\[3mm] \widetilde{\pi}(y) = \dfrac{\pi_{00}(1 - p_{10})(yp_{02}^0 + 1)}{p_{-1,-1} - p_{-1,1}y}. \end{cases}$$

In fact (4.5.2) gives the simplest example allowing reduce the problem to the case $f \equiv 1$. To find less evident cases (for groups of order 4), when

$$N(f) = f f_\delta = 1,$$

one must check that

$$A = A_\xi, \text{ with } A = \frac{q q_\delta}{\widetilde{q}\, \widetilde{q}_\delta}. \tag{4.5.5}$$

But A can be rewritten as

$$A = \frac{a(x)y + b(x)}{c(x)y + d(x)} \mod Q(x,y),$$

for some $a, b, c, d \in \mathbb{C}$, and we get the following necessary and sufficient condition for (4.5.5) to hold:

$$a(x)d(x) - b(x)c(x) \equiv 0.$$

It is immediate to see that this condition yields a set of polynomial equations in the parameters. A particular case arises when $q q_\delta = (q q_\delta)_\xi$, which means that

$$a(x) \equiv 0.$$

The number of corresponding equations is then exactly 3. Setting $\alpha = x\eta(x)$, we have

$$\begin{cases} p'_{11}p'_{-1,0} = p'_{10}p'_{-1,1}, \\[3mm] -p'_{11} + \dfrac{p'_{01}p'_{-1,0}}{\alpha} = p'_{10}p'_{01} - \dfrac{p'_{-1,1}}{\alpha}, \\[3mm] p'_{01}p'_{10}\alpha - p'_{-1,1} = p'_{-1,0}p'_{01} - p'_{11}\alpha. \end{cases} \tag{4.5.6}$$

Similar equations can be obtained when

$$\widetilde{q}\, \widetilde{q}_\delta = (\widetilde{q}\, \widetilde{q}_\delta)_\xi.$$

We see that polynomial equations even more complicated than (4.5.6) appear and it is not the right way of tackling the problem.

4.5.1.3 One Parameter Families A more interesting example was found in
[26]. Let us consider, in continuous time, a *simple* random walk such that

$$
\begin{cases}
\dfrac{Q(x,y)}{xy} = p_{10}(1-x) + p_{01}(1-y) + p_{-1,0}\left(1 - \dfrac{1}{x}\right) + p_{0,-1}\left(1 - \dfrac{1}{y}\right), \\[2mm]
\dfrac{q(x,y)}{xy} = t\left(\dfrac{1}{x} - 1\right) + p_{0,-1}\left(1 - \dfrac{1}{y}\right), \\[2mm]
\dfrac{\tilde{q}(x,y)}{xy} = t\left(1 - \dfrac{1}{y}\right) + p_{-1,0}\left(1 - \dfrac{1}{x}\right).
\end{cases}
$$

We note that the equations

$$
\begin{cases}
q = 0 \Leftrightarrow t = \dfrac{p_{0,-1}x(1-y)}{y(1-x)}, \\[3mm]
\tilde{q}_\delta = 0 \Leftrightarrow t = \dfrac{p_{-1,0}\beta(x-\alpha)}{\alpha(\beta - y)},
\end{cases}
$$

where

$$
\alpha = x\eta(x), \qquad \beta = y\xi(y),
$$

provide a parametrization of the curve $Q(x,y) = 0$ by means of the parameter
t. Indeed the following equivalence holds

$$
\frac{p_{0,-1}x(1-y)}{y(1-x)} = \frac{p_{-1,0}\beta(x-\alpha)}{\alpha(\beta - y)} \Leftrightarrow Q(x,y) = 0.
$$

Here $N(f) = f f_\delta = 1$. Thus exists $c \in \mathbb{C}(x)$, such that $c = f c_\delta$. On the other
hand, if condition (4.2.17) holds, which writes

$$
\psi f_\delta + \psi_\delta = 0,
$$

then the generating functions π and $\tilde{\pi}$ are rational, and given by

$$
\pi(x) = \frac{\pi(0)}{1 - \rho_1 x}, \qquad \tilde{\pi}(y) = \frac{\tilde{\pi}(0)}{1 - \rho_2 y}, \tag{4.5.7}
$$

where ρ_1 and ρ_2 are the positive roots of second degree equations in the parameter space.

4.5.1.4 Two Typical Situations Here we want to show that the above examples are typical in the following sense: a very large number of reasonably
simple many-parameter families of rational probabilistic solutions exist, in the
case $N(f) = 1$, but there is only a finite number of one-parameter families for
which $N(f) \neq 1$. First of all, we shall give immediate necessary conditions for
rationality, when π and $\tilde{\pi}$ have no poles in the unit disk.

Definition 4.5.1 *Let a set X and a group G of its one-to-one mappings $g :
X \to X$. For an arbitrary but fixed $x \in X$, the set of all $g(x)$, $g \in G$, is called
the orbit of x under G.*

Lemma 4.5.2   *Consider a simple random walk and let* $\mathcal{H} \overset{def}{=} \{1, \xi, \eta, \delta\}$. *Then, for* π, $\tilde{\pi}$ *to be rational, it is necessary that there exists a point* $s_0 \in \mathbf{S}$, *such that* $x(s)$ *and* $y(s)$ *be real at all points* $s \in \mathcal{H}s_0 = \{hs_0 : h \in \mathcal{H}\}$, *the orbit of* s_0, *and that at least one of the following conditions be satisfied:*

1. $q(h_1 s_0) = q(h_2 s_0)$, *for some* $h_1 \neq h_2$, $h_1, h_2 \in \mathcal{H}$;

2. $\tilde{q}(h_1 s_0) = \tilde{q}(h_2 s_0)$, *for some* $h_1 \neq h_2$, $h_1, h_2 \in \mathcal{H}$;

3. $q(h_1 s_0) = \tilde{q}(h_2 s_0)$, *for some* $h_1, h_2 \in \mathcal{H}$;

4. *one of the functions* q *or* $\tilde{q}$ *vanishes at a point* $s \in \mathcal{H}s_0$, *which is a fixed point of* ξ *or* η.

∎

Before proving this lemma, we shall give hereafter some properties of the orbits and also prove two auxiliary lemmas 4.5.3, 4.5.4. Two cases will be of special interest:

$$\begin{cases} X = \mathbf{S}, & G = \mathcal{H}, \quad \text{and} \\ X = \Omega, & G = \tilde{\mathcal{H}}, \quad \text{the lifting of } \mathcal{H} \text{ onto } \Omega. \end{cases}$$

In the second case, it will be convenient to deal with $\Omega/_{\{n\omega_1\}}$, since all functions have the period to ω_1.

We are now in a position to quote three useful properties **R1, R2, R3**.

R1 The group $\mathcal{H}$, is finite if, and only if, all the orbits $\mathcal{H}s$, $s \in \mathbf{S}$, are finite.

R2   Setting $I \overset{def}{=} h_x^{-1}(\mathcal{D}) \cap h_y^{-1}(\mathcal{D})$, there are situations where

$$\bar{I} \cap \xi\bar{I} = \bar{I} \cap \eta\bar{I} = \emptyset, \tag{4.5.8}$$

e.g. for the groups of order 6 shown in figure 4.5.2, when $M_x < 0$, $M_y < 0$. In

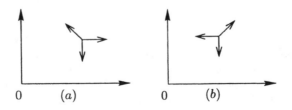

$$0 \qquad (a) \qquad\qquad 0 \qquad (b)$$

Fig. 4.5.2.

fact, if e.g. $\bar{I} \cap \eta\bar{I} \neq \emptyset$, then Γ_0 and $\eta(\Gamma_0)$, have a non empty intersection. But on Γ_0, $|x| = |y| = 1$ and we get a contradiction, since, for instance in figure 4.5.2-(b)

$$|\eta x| = \left| \frac{p_{-1,0}}{p_{11}xy} \right| = \frac{p_{-1,0}}{p_{11}} > 1.$$

Suppose that (4.5.8) holds. Then, for any $\omega \in \Delta_0$, $\eta\xi\eta(\omega) \notin \Delta_0$,

$$\xi\eta\xi(\omega) \notin \Delta_0.$$

R3   Any orbit $\tilde{\mathcal{H}}\omega$ in Ω can be mapped in a one-to-one manner onto the set $\mathbb{Z}$. More exactly we can order $\tilde{\mathcal{H}}\omega$ in the following way:

$$\cdots < \xi\eta(\omega) < \eta(\omega) < \omega < \xi(\omega) < \eta\xi(\omega) < \xi\eta\xi(\omega) < \cdots ,$$

where the element ω has number 0 and, in case of equalities, we consider e.g. a and $\xi(a)$ are formally different. Thus we can call *interval of the orbit* any inverse of some interval $[m, n] \in \mathbb{Z}$, and the notion of *interval* does not depend on the choice of the point ω on the orbit.

Lemma 4.5.3   *For any ω, the intersection $\{\tilde{\mathcal{H}}\omega\} \cap \overline{\Delta}_0$ is an interval, both ends of which belong to $\overline{\Delta}_0 \backslash \overline{I}$. All other points of this interval (if there are some) belong to $\overline{I}$.*   ∎

Proof.   Let $\omega \in \overline{I}$. Then $\xi(\omega) \in \overline{\mathcal{D}}_1 \subset \overline{\Delta}_0$. If now $\xi(\omega) \in \overline{\mathcal{D}}_1 \backslash \overline{I}$, then $\eta\xi(\omega)$, $\xi\eta\xi(\omega)$, ... do not belong to $\overline{\Delta}_0$, as $\omega, \eta(\omega) \in \overline{\Delta}_0$ if, and only if, $\omega, \eta(\omega) \in \{\omega : |y(\omega)| \leq 1\} \cap \overline{\Delta}_0$. If $\xi(\omega) \in \overline{I}$, then similar arguments hold for $\omega_1 = \xi(\omega)$, $\eta(\omega_1)$, etc.   ∎

From lemma 4.5.3, one can derive similar properties for the orbit

$$\{\tilde{\mathcal{H}}_0\omega\} = \{\omega + n\omega_3\}.$$

In section 4.6, it will be shown, for the group of order 6 corresponding to figure 4.5.2-(b), that

$$\omega_3 = \frac{2}{3}\omega_2.$$

4.5.1.5 Ergodicity Conditions We can explain in an elementary way the analytical nature of the ergodicity conditions. The functions $q(x, y)$ and $\tilde{q}(x, y)$ are supposed both to have a zero at the point $s \in \mathbf{S}$, $x(s) = y(s) = 1$.

Lemma 4.5.4   *The function $q(x, y)$ [resp. $\tilde{q}(x, y)$] has at the point $s \in \mathbf{S}$, $x(s) = y(s) = 1$, a zero of at least second order if, and only if,*

$$M_x M_y' = M_y M_x' \tag{4.5.9}$$

$$[resp. \ M_y M_x'' = M_x M_y''] \tag{4.5.10}$$

∎

Proof.   Let $M_y \neq 0$. Then $x(s) - 1$ has a zero of first order at the point $(1,1) \in \mathbf{S}$, so that q has, at the same point, a zero of order not less than two if, and only if,

$$\frac{dp'(x,y)}{dx}\bigg|_{x=y=1} = 0. \tag{4.5.11}$$

But, for $x = y = 1$,

$$\frac{dy}{dx} = -\frac{p_x(x,y)}{p_y(x,y)} = -\frac{M_x}{M_y}$$

and

$$\frac{dp(x,y)}{dx} = \frac{dp'(x,y)}{dy}\frac{dy}{dx} + \frac{dp'(x,y)}{dx} = -M_y'\frac{M_x}{M_y} + M_x',$$

which shows that (4.5.11) is equivalent to (4.5.9), remembering that the functions $p(x,y)$, $p'(x,y)$, $p''(x,y)$ have been defined in chapter 1.

In the case $M_y = 0$, $M_x \neq 0$, one should also take into account the quantity $\dfrac{dx}{dy}$.

It will be shown now that, under conditions (4.5.9) or (4.5.10), the system can never be ergodic. In fact, the left hand side of the fundamental equation

$$q\pi + \widetilde{q}\widetilde{\pi} = -q_0\pi_{00},$$

does have zero of at least order two at the point (1,1). But it is always possible to choose q_0 (which has no effect on the ergodicity) with a zero of first order at (1,1). The proof of lemma 4.5.4 is concluded.   ∎

The non ergodicity in the other cases can be proved along the same lines. In fact, there is one *travelling* zero of q, which coincides with (1,1) when (4.5.9) holds (a similar travelling zero exists for $\widetilde{q}$, up to a change of parameters) and influences the ergodicity conditiond. But it is necessary to have information about other zeros, to make sure that π and $\widetilde{\pi}$ have no poles in the unit circle.

4.5.1.6 Proof of Lemma 4.5.2 The power series representing $\pi(x)$ has positive coefficients. Thus if $\pi(x)$ is rational, then it has a positive pole at some point $x_0 > 1$, by the Hadamard-Pringsheim theorem.

Taking a point $s_0 \in \mathbf{S}$ such that $x(s_0) = x_0$, we shall first show that the orbit $\mathcal{H}s_0$ is real, i.e. all values x and y on this orbit are real. For this it is sufficient to prove that when $y(s_0)$ and $y(\xi(s_0))$ are complex conjugate (not real) leads to contradiction. Since any orbit contains not more than 4 points and intersects with Δ, then either

$$\text{(a)} \quad |y(s_0)| \leq 1 \qquad \text{or} \quad \text{(b)} \quad |x(\eta(s_0))| \leq 1. \tag{4.5.12}$$

As for (a), one must have $q(s_0) = 0$, since otherwise $\widetilde{\pi}(s_0) = \infty$, which is impossible. But q is of first degree in y and cannot be zero for non real y.

In case (b), again $q(s_0) \neq 0$. Hence $\widetilde{\pi}$ has a pole at s_0, so that

$$\widetilde{q}(\eta\xi(x_0)) = \widetilde{q}(\eta(s_0)) = 0.$$

But $\widetilde{q}$ is equal mod Q to a polynomial of first degree in y. Hence the case (b) is again impossible since $x(\eta(s_0)) > 0$.

Continuing with the puzzle, assume now first $|y(\eta(s_0))| < 1$, for some $\eta \in \mathcal{H}$. We have $y(\eta(s_0)) = y(s_0)$ and $y(\eta\xi(s_0)) = y(\xi(s_0))$. Hence, either $|y(s_0)| < 1$ or $|y(\xi(s_0))| < 1$. But since π (and also $\widetilde{\pi}$) is rational, any of these points is suitable for our purpose. Take for instance $|y(s_0)| < 1$. But then $q(s_0) = 0$, because otherwise $\widetilde{\pi}$ would have a pole. Hence we are left with two possible situations:

$$\text{(i)} \quad |x(\eta(s_0))| < 1 \qquad \text{or} \quad \text{(ii)} \quad |x(\eta(s_0))| > 1. \tag{4.5.13}$$

If (i) takes place together with $y(\xi(s_0)) \neq 0$, then $\widetilde{\pi}$ has a pole at $s = \xi(s_0)$ and consequently $\widetilde{q}(\eta\xi(s_0)) = 0$, since otherwise π would have a pole at $\xi\eta(s_0)$, which is impossible due to the inequality $|x(\xi\eta(s_0))| < 1$.

In the case (ii), it can be shown along the sames lines that, if $q(\xi(s_0)) \neq 0$ and $\widetilde{q}(\eta\xi(s_0)) \neq 0$, then π must have a pole at the point $\xi\eta(s_0)$. Then $q(\eta(s_0)) = 0$, since otherwise $\widetilde{\pi}(\eta(s_0)) = \infty$.

Let $|y(s)| > 1$, for all points s of the orbit. Then $|x(\eta(s_0))| < 1$, as any orbit intersects with Δ. If both $q(s_0) \neq 0$ and $q(\xi(s_0)) \neq 0$, then $\widetilde{\pi}$ has poles at s_0 and $\xi(s_0)$. Thus $\widetilde{q}(\eta(s_0)) = \widetilde{q}(\eta\xi(s_0)) = 0$, still by the same arguments. If only one of the quantities $q(s_0)$ or $q(\xi(s_0))$ vanishes, then $\widetilde{\pi}$ has a pole at the corresponding point and then $\widetilde{q}(hs) = 0$.

Assume now that in $\mathcal{H}s_0$ there exists a fixed point of ξ or η. It is then easy to show that $\mathcal{H}s_0$ consists of exactly two points, one being a fixed point of ξ, the other one a fixed point of η. If for instance $s_0 = \xi(s_0)$, then either $q(s_0) = 0$ or $\widetilde{q}(s_0) = 0$. Otherwise, either $\pi(x)$ has a pole at the second point of the orbit, or $\widetilde{\pi}(s_0) = \infty$. But this is impossible since $(\mathcal{H}s_0) \cap \Delta \neq \emptyset$.

Take now the orbit of the point $y(s) = x(s) = 1$ and let

$$\pi\left(\frac{p_{-1,0}}{p_{10}}\right) = \widetilde{\pi}\left(\frac{p_{0,-1}}{p_{01}}\right) = \infty,$$

so that

$$q\left(\frac{p_{-1,0}}{p_{10}}, 1\right) = \widetilde{q}\left(1, \frac{p_{0,-1}}{p_{01}}\right) = 0.$$

But, if $\pi\left(\dfrac{p_{-1,0}}{p_{10}}\right)$ is finite, then π must have a pole at the automorphic point, and similarly for $\widetilde{\pi}$. Therefore the above arguments could be used and the proof of lemma 4.5.1 is concluded. ∎

Corollary 4.5.5 *For given p_{ij}'s, the set of parameters $\{p'_{ij}, p''_{ij}\}$ which produce rational π and $\widetilde{\pi}$ belong to a hypersurface in the space of the parameters. Incidentally, it is worth noting that this result can be proved in several different ways.*

From lemma 4.5.2, one sees that two situations prevail:

(i) $q \equiv q_h$ or $\widetilde{q} \equiv \widetilde{q}_h$ or $q \equiv \widetilde{q}_h$, for some $h \in \mathcal{H}$.

This case includes, in particular, reversibility and also (4.5.2). Here many-parameter families exist which produce rational solutions.

(ii) There is a one-parameter algebraic family $q(x, y; c)$, $\widetilde{q}(x, y; c)$ [with a clear notational device], such that e.g. q_h and $\widetilde{q}_h$ have a common zero $(x(c), y(c))$, for any value of the parameter c. Then after eliminating c from the two equations $q_h = \widetilde{q}_h = 0$, we get an algebraic curve, which either coincides with Q or is a degenerate one-point set. The dependency with respect to another parameter (if any) can be only trivial. As remarked earlier, more sophisticated methods might well exists for this problem, which seems to be of a computational nature.

4.6 An Example of Algebraic Solution by Flatto and Hahn

This is the first example of algebraic non rational solution, which was encountered and explicitly solved. Moreover it corresponds to a very natural Markovian queueing model, with two servers supplied with parallel arrivals. The analysis of this model gives rise to a random walk in $\mathbb{Z}_+^2$, shown in figure 4.6.1 and having the following parameters:

$$\begin{cases} Q(x, y) &= \dfrac{\left[x^2 y^2 - (1 + \alpha + \beta)xy + \beta x + \alpha y\right]}{1 + \alpha + \beta}, \\[2ex] q &= \dfrac{yx^2 - (\alpha + 1)x + \alpha}{1 + \alpha}, \\[2ex] \widetilde{q} &= \dfrac{xy^2 - (\beta + 1)y + \beta}{1 + \beta}, \\[2ex] q_0 &= xy - 1. \end{cases}$$

Setting

$$\pi_1(x) = \frac{\beta x[x\pi(x) + \pi_{00}]}{1 - x}, \qquad \widetilde{\pi}_1(y) = \frac{\alpha y[y\widetilde{\pi}(y) + \pi_{00}]}{1 - y},$$

the fundamental equation takes the form

$$\pi_1(x) + \widetilde{\pi}_1(y) = 0, \quad \text{on } Q(x, y) = 0. \tag{4.6.1}$$

From section 4.3, it appears that the solution of (4.6.1) (when it exists) is necessarily algebraic, since one can easily check that all conditions of theorem 4.3.1 are satisfied. To find this solution, we shall proceed in constructing the algebraic extension $\mathbb{F}_2$ of $\mathbb{C}_Q(x, y)$, which is defined by the equation

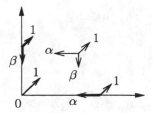

Fig. 4.6.1.

$$v^2 = x_3 - x, \qquad (4.6.2)$$

where x_3 is the branch point of $y(x)$ outside the unit circle (here $x_4 = \infty$). The group $\mathcal{H}$ is of order 6 and, in addition, π_1 and $\tilde{\pi}_1$ belong to

$$\mathbb{C}_{\delta,\xi} = \mathbb{C}_{\delta,\eta} = \mathbb{C}_\delta \cap \mathbb{C}_\xi \subset \mathbb{C}_\delta \subset \mathbb{F}_2.$$

Lemma 4.6.1 *For the random walk shown in figure 4.6.1, we have*

$$\omega_3 = \frac{2\omega_2}{3}. \qquad (4.6.3)$$

∎

Proof. Here

$$\begin{cases} 0 < x_1 < x_2 < 1 < x_3, \quad x_4 = \infty, \\ \alpha \le x_3 \le \beta \le y_3, \quad y_j = \dfrac{\beta x_j}{\alpha}, \quad 1 \le j \le 3. \end{cases}$$

We note immediately that $\omega_3 \in \left\{ \dfrac{\omega_2}{3}, \dfrac{2\omega_2}{3} \right\}$. According to the uniformization (3.3.4), we have

$$x(0) = \infty, \quad y(0) = 0, \quad \eta(\omega) = \omega_2 + \omega_3 - \omega, \quad \forall \omega \in \mathbb{C}_\omega.$$

Writing for any automorphism $h : (x(\omega), y(\omega)) \to (x(h(\omega)), y(h(\omega)))$, we obtain

$$\eta(\infty, 0) = (0, 0),$$

and, still by (3.3.3),

$$z(\eta(0)) = z(\omega_2 + \omega_3) = \alpha > 0,$$

which implies $\wp'(\omega_2 + \omega_3) > 0$. Using standard properties of the Weierstrass function $\wp$, it follows that $\omega_2 \le 2\omega_3$ and (4.6.3) holds. The proof of the lemma is complete. ∎

Lemma 4.6.2

$$\pi \in \mathbb{F}_2. \tag{4.6.4}$$

∎

Proof.   On $\mathbb{C}_\omega$, π_1 is elliptic, with periods ω_1, ω_3, and satisfies

$$(\pi_1)_\xi = \pi_1, \quad \text{or} \quad \pi_1(\omega_2 - \omega) = \pi_1(\omega).$$

Moreover, by (4.6.3), π_1 admits also the period $2\omega_2$. Consider now the function v, defined by (4.6.2), which is algebraic of x, elliptic with periods $\omega_1, 2\omega_2$, and enjoys the following properties:

- 2 simple poles at $\omega = 0$, $\omega = \omega_2$, with respective residues 1 and -1;

- 2 simple zeros at $\omega = \dfrac{\omega_1}{2}$, $\omega = \dfrac{\omega_1}{2} + \omega_2$, since x_3 is a branch-point.

The residues at 0 and ω_2 having opposite signs, v does not have the period ω_2.

Upon setting for a while $\pi^*(\omega) \stackrel{\text{def}}{=} \pi_1(\omega + \omega_2/2)$, one sees immediately that π^* is an even function, elliptic with periods $\omega_1, 2\omega_2$. Consequently, $\pi^*(\omega)$ is a rational function of $\wp(\omega; \omega_1, 2\omega_2)$.

Exactly the same is true for the function $v^*(\omega) \stackrel{\text{def}}{=} v(\omega + \omega_2/2)$. In addition, v^* is of order 2, since it has 2 poles and 2 zeros in the parallelogram $[0, 2\omega_2[\times [0, \omega_1[$. Thus $v^*(\omega)$ is fractional linear transform of $\wp(\omega; \omega_1, 2\omega_2)$, and hence π_1 (and consequently π) is also a rational function of v. The proof of the lemma is terminated.   ∎

In fact we have proved that π belongs to the extension of $\mathbb{C}(x)$, of which v is a generator. This extension is nothing else, but $\mathbb{C}(v)$.

The final step is to find the rational function mentioned in the proof of lemma 4.6.2. At that moment, we could proceed as done in [31]. Instead, we will use the general machinery developped in this chapter.

First some properties of the branches shown in section 5.3 ensure that

$$\pi_1(x) \sim C/\sqrt{x_3 - x}, \quad \text{for } x \to \infty.$$

Using now (4.6.2), corollary 4.3.4 with $k = 2$, $n = 3$, we deduce easily that Δ_0 (coming in this corollary) is in fact a fractional linear transform. The function Δ_0 is precisely computed from its values at 3 specific points, since

$$\pi_1(0) = \pi_1(\infty) = 0, \quad \pi_1(1) = \infty.$$

This yields

$$x\pi(x) + \pi_{00} = \frac{\beta - 1}{\beta} \frac{g(x)}{g(1)},$$

with

$$g(x) = \frac{\sqrt{x_3 - x} + \sqrt{x_3 - 1}}{\left(\sqrt{x_3 - x} + \sqrt{x_3 - \alpha/\beta}\right)\left(\sqrt{x_3 - x} - \sqrt{x_3 - \alpha}\right)}.$$

Similarly

$$y\widetilde{\pi}(y) + \pi_{00} = \frac{\alpha - 1}{\alpha} \frac{\widetilde{g}(y)}{\widetilde{g}(1)},$$

with

$$\widetilde{g}(y) = \frac{\sqrt{y_3 - y} + \sqrt{y_3 - 1}}{\left(\sqrt{y_3 - y} + \sqrt{y_3 - \beta/\alpha}\right)\left(\sqrt{y_3 - y} + \sqrt{y_3 - \beta}\right)}.$$

For further computational details, the reader can consult [30, 31].

4.7 Two Queues in Tandem

This random walk is shown in figure 4.5.2-(b). In the simplest case, when the parameters are continuous on the boundaries, the stationary distribution is given by the famous *product form of Jackson's networks*, analogous to (4.5.7), which involves very simple rational generating functions. We want to emphasize that rational solutions can exist even when the group for these networks is infinite, although we could not exactly immerse this phenomenon in the theory proposed in this book.

The time-dependent behaviour of this system, analyzed by Blanc [7], is much less elementary. It reduces, up to an additional Laplace tranform, to the analysis of our fundamental functional equation. The group $\mathcal{H}$ is still of order 6, with $N(f) = 1$, but the corresponding trace is not zero: this shows immediately that the solution cannot be algebraic.

5. Solution in the Case of an Arbitrary Group

In chapter 4, the analysis was based on specific derivations (a closure property in some sense) rendered possible by the finiteness of the order of the group. Hereafter, we shall obtain the complete solution when the order of the group of the random walk is arbitrary, i.e. possibly infinite. The main idea consists in the reduction to a factorization problem on a curve in the complex plane. Generally one comes up first with integral equations and, in a second step, with explicit integral forms by means of Weierstrass functions.

In theorem 3.2.2, it has been proved that π [resp. $\tilde{\pi}$] can be continued as a meromorphic function to the whole complex plane, cut along $[x_3x_4]$ (resp. $[y_3y_4]$). This result was presented as a consequence of analytic continuations on the Riemann surface $\mathbf{S}$ and on the universal covering, by projection and lifting operations. The derivation of theorem 3.2.2 was heavily relying on specific properties of the algebraic curve $Q(x, y) = 0$, given in lemma 2.3.4.

In this chapter, closed form solutions for the functions π and $\tilde{\pi}$ will be produced, by reduction to *boundary value problem* (BVP) of Riemann-Hilbert type in the complex plane, following the approach originally proposed in [26].

5.1 Informal Reduction to a Riemann-Hilbert-Carleman BVP

We shall in fact solve the problem under conditions slightly more general than the ones which produced equation (1.3.6). Referring to equation (1.3.4) for the stationary probabilities of a piecewise homogeneous random walk, and using the notation of section 1.3, we suppose that the two-dimensional lattice is partitioned into $M + L + 2$ classes S_r, such that $\displaystyle\bigcup_{r=1}^{M+L+2} S_r = \mathbb{Z}_+^2$ and

$$
S_r = \begin{cases}
S_i' &= \{(i, 0)\}, \quad i = 1, \dots, L - 1 \\
S' &= \{(i, 0),\ i \geq L\}, \\
S_j'' &= \{(0, j)\}, \quad j = 1, \dots, M - 1 \\
S'' &= \{(0, j),\ j \geq M\}, \\
S &= \{(i, j)\ ;\ i \geq 1,\ j \geq 1\}, \\
S_0 &= \{(0, 0)\}.
\end{cases}
$$

Incidentally, the generating functions P_r of the jumps in the region r will be, for convenience, denoted by $P_{k\ell}$, for $(k, \ell) \in S_r$, and they depend only on r. Then equation (1.3.4) becomes

$$\boxed{-Q(x, y)\pi(x, y) = q(x, y)\pi(x) + \widetilde{q}(x, y)\widetilde{\pi}(y) + \pi_0(x, y),} \qquad (5.1.1)$$

where

$$\begin{cases} \pi(x, y) = \sum_{i,j \geq 1} \pi_{ij} x^{i-1} y^{j-1}, \\[2mm] \pi(x) = \sum_{i \geq L} \pi_{i0} x^{i-1}, \quad \widetilde{\pi}(y) = \sum_{j \geq M} \pi_{0j} y^{j-M}, \\[2mm] q(x, y) = x^L \left(\sum_{i \geq -L, j \geq 0} p'_{ij} x^i y^j - 1 \right) \equiv x^L (P_{L0}(x, y) - 1), \\[2mm] \widetilde{q}(x, y) = y^M \left(\sum_{i \geq 0, j \geq -M} p''_{ij} x^i y^j - 1 \right) \equiv y^M (P_{0M}(x, y) - 1), \\[2mm] \pi_0(x, y) = \sum_{i=1}^{L-1} \pi_{i0} x^i (P_{i0}(x, y) - 1) + \sum_{j=1}^{M-1} \pi_{0j} y^j (P_{0j}(x, y) - 1) \\[2mm] \qquad\qquad + \pi_{00}(P_{00}(x, y) - 1), \end{cases} \qquad (5.1.2)$$

$Q(x, y)$ being given in (1.3.5).

The reader will have noticed that the above partitioning implies that no assumption is made about the boundedness of the upward jumps on the axes, neither at $(0, 0)$. In addition, the downward jumps on the x [resp. y] axis are bounded by L [resp. M], where L and M are arbitrary finite integers.

According to sections 2.2 and 2.3, the fundamental equation (5.1.1) can be restricted to the algebraic curve $\mathcal{A}$ defined by $Q(x, y) = 0, (x, y) \in \mathbb{C}^2$. With the notations of chapter 2, this yields

$$q(X_0(y), y)\pi(X_0(y)) + \widetilde{q}(X_0(y), y)\widetilde{\pi}(y) + \pi_0(X_0(y), y) = 0, \ y \in \mathbb{C}. \qquad (5.1.3)$$

We make the additional assumption that the given generating functions P_{i0} and P_{0j} in (5.1.2) have *suitable* analytic continuations, in a sense which will be rendered more precise in section 5.4.

Letting now y tend successively to the upper and lower edge of the slit $[y_1 y_2]$, using the fact that $\widetilde{\pi}$ is holomorphic in $\mathcal{D}$ and in particular on $[y_1 y_2]$, we can eliminate $\widetilde{\pi}$ in (5.1.3) to get

$$\pi(X_0(y))f(X_0(y), y) - \pi(X_1(y))f(X_1(y), y) = h(y), \text{ for } y \in [y_1 y_2], \qquad (5.1.4)$$

or, anticipating slightly on the results of section 5.3 about the functions $X_0(y)$ and $Y_0(x)$,

$$\boxed{\pi(t)A(t) - \pi(\alpha(t))A(\alpha(t)) = g(t), \quad t \in \mathcal{M},}$$
(5.1.5)

where

$$
\begin{cases}
f(x,y) & = \dfrac{q(x,y)}{\widetilde{q}(x,y)}, \quad A(x) = f(x, Y_0(x)), \\[2mm]
h(y) & = \dfrac{\pi_0(X_1(y),y)}{\widetilde{q}(X_1(y),y)} - \dfrac{\pi_0(X_0(y),y)}{\widetilde{q}(X_0(y),y)}, \quad g(x) = h(Y_0(x)), \\[2mm]
\alpha(x) & = \overline{x},
\end{cases}
$$
(5.1.6)

and $\mathcal{M}$ is a simple closed contour $X_0[y_1 y_2]$. It turns out that the determination of π, meromorphic in the interior of the domain bounded by $\mathcal{M}$, is possible and is equivalent to solving a BVP of Riemann-Hilbert-Carleman type, on the coutour $\mathcal{M}$ in the complex plane. Obviously, as aforementioned, we have supposed that the given functions P_r have meromorphic continuations in the domain bounded by $\mathcal{M}$.

The next section is devoted to a general survey of the basic theoretical results for BVP's in $\mathbb{C}$.

5.2 Introduction to BVP in the Complex Plane

5.2.1 A Bit of History

These last fifty years, a huge literature, mainly originating from the still so-called soviet school, has been devoted to the extensive study of BVP such they arise in the theory of elasticity, hydromechanics and other fields of mathematical physics.

In fact, these problems appeared *in embryo* at the end of the last century. At that time important results were obtained in two directions:

- First, the interpretation of the real and imaginary parts of Cauchy-type integrals as the potential of simple and double layers, respectively (Harnack, e.g. [61]).

- Secondly, the characterization of the limiting values of Cauchy-type integrals on the contour of integration (Sokhotski and Plemelj, e.g. [67, 61]), involving the so-called *principal* value of singular integrals.

Nevertheless, B. Riemann was probably the first in his dissertation to have mentioned the following general problem:

Find a function which is holomorphic in some domain $\mathcal{D}$ and continuous on the boundary of $\mathcal{D}$, for a given relation between the limiting values of its real and imaginary parts.

This problem was studied by Hilbert in 1905, who gave a partial answer, by reduction to integral equations. Then H. Poincaré in his *Leçons de Mécanique Céleste* [68] came up with a boundary value problem for harmonic functions, which led him to study singular integral equations in more details.

In this overall presentation, we follow essentially Muskhelischvili [61], Gakhov [35] and Litvintchuk [44].

Let $\mathcal{L}$ be a simple smooth line or curve in the complex plane, i.e. an arc (open or closed) with a continuously varying tangent (see [61]). A function f will be said to satisfy the *Hölder condition* on the curve $\mathcal{L}$ if, for any two points t_1, t_2 of $\mathcal{L}$,

$$|f(t_2) - f(t_1)| \leq A|t_2 - t_1|^\mu, \tag{5.2.1}$$

where A and μ are positive constants. Clearly for $\mu > 1$ the above condition (5.2.1) implies $f = \text{constant}$. Hence we only consider the interesting situation $0 < \mu \leq 1$.

When (5.2.1) holds, we shall write, according to a well established tradition, $\varphi \in \mathbb{H}_\mu(\mathcal{L})$, for all φ bounded on $\mathcal{L}$. It is also important to note that $\mathbb{H}_\mu(\mathcal{L})$ can be endowed with the norm

$$\|\varphi\| = \sup_{t \in \mathcal{L}} |\varphi(t)| + \sup_{s,t \in \mathcal{L}} \frac{|\varphi(s) - \varphi(t)|}{|s - t|^\mu}.$$

5.2.2 The Sokhotski-Plemelj Formulae

These formulae were apparently firstly discovered by Sokhotski in 1873 (although not completely rigourously proved, as claimed in Gakhov [35]), who investigated the limiting behavior of Cauchy-type integrals on the contour $\mathcal{L}$. Nevertheless, Muskhelishvili refers to them as *Plemelj formulae*, from the work of Plemelj [67] in 1908, since the arguments were sufficiently rigourous.

Let $\varphi \in \mathbb{H}_\mu(\mathcal{L})$. We assume that a positive direction has been chosen on $\mathcal{L}$ (in the case of a closed contour, positive means counter-clockwise). The contour $\mathcal{L}$ is not necessarily bounded. Then the following classical theorem holds (see for example [61], [35]):

Theorem 5.2.1   *The function*

$$\Phi(z) = \frac{1}{2i\pi} \int_{\mathcal{L}} \frac{\varphi(t)dt}{t - z} \, , z \notin \mathcal{L}. \tag{5.2.2}$$

is continuous on $\mathcal{L}$ from the left and from the right, with the exception of the ends. Moreover the corresponding limiting values, denoted respectively by Φ^+

and Φ^-, are in the class $\mathbb{H}_\mu(\mathcal{L})$, and they satisfy the so-called Sokhotski-Plemelj formulae, for $t \in \mathcal{L}$,

$$
\begin{cases}
\Phi^+(t) & = & \dfrac{1}{2}\varphi(t) + \dfrac{1}{2i\pi}\displaystyle\int_{\mathcal{L}} \dfrac{\varphi(s)ds}{s-t}, \\[3mm]
\Phi^-(t) & = & -\dfrac{1}{2}\varphi(t) + \dfrac{1}{2i\pi}\displaystyle\int_{\mathcal{L}} \dfrac{\varphi(s)ds}{s-t},
\end{cases}
\tag{5.2.3}
$$

where the integrals are understood in the sense of Cauchy-principal value. Substracting and adding formulae (5.2.3), one obtains two other equivalent and fundamental formulae, for $t \in \mathcal{L}$, namely

$$
\begin{cases}
\Phi^+(t) - \Phi^-(t) & = & \varphi(t), & \tag{5.2.4} \\[3mm]
\Phi^+(t) + \Phi^-(t) & = & \dfrac{1}{i\pi}\displaystyle\int_{L} \dfrac{\varphi(s)ds}{s-t}. & \tag{5.2.5}
\end{cases}
$$

In fact, (5.2.4) can be shown to be valid for $\varphi(t)$ simply continuous on $\mathcal{L}$, provided some restriction are put on the way of approaching $\mathcal{L}$ (which should not be, roughly said, too tangential); on the other hand, cusp points and corner points, when they exist, compel to modify (5.2.3) (see e.g. [61]). ∎

5.2.3 The Riemann Boundary Value Problem for a Closed Contour

Let $\mathcal{L}$ be now a closed contour without self-intersection. It does divide the complex plane into two parts: the interior domain within $\mathcal{L}$, denoted by $\mathcal{L}^+$ and the exterior domain (the complementary of $\mathcal{L}^+$) denoted by $\mathcal{L}^-$ (when traversing $\mathcal{L}$ in the positive direction, $\mathcal{L}^+$ remains on the left).

We shall say that a function Φ is *sectionally holomorphic* if Φ is holomorphic in every finite region of the plan, except on $\mathcal{L}$ where it has left and right limits Φ^+ and Φ^-. Moreover this function Φ will be said to have a *finite degree at infinity*, if the only singularity at infinity is a pole of finite order. It will be convenient to call Φ sectionally holomorphic, whenever $\Phi(\infty) = $ Constant.

The *Riemann* BVP can be stated as follows: *Find a sectionally holomorphic function Φ of finite degree at infinity, under the boundary condition on $\mathcal{L}$*

$$
\boxed{\Phi^+(t) = G(t)\Phi^-(t) + g(t), \quad t \in \mathcal{L},}
\tag{5.2.6}
$$

where $G \in \mathbb{C}(\mathcal{L})$, the space of continuous functions on $\mathcal{L}$, and $g \in L_p(\mathcal{L})$. A priori g and G are only defined on $\mathcal{L}$.

For our purpose, we shall only consider the problem for $G, g \in \mathbb{H}_\mu(\mathcal{L})$ and also assume that G does not vanish on $\mathcal{L}$.

Notation   Introduce the following important quantity

$$\chi = \mathcal{I}nd[G]_L = \frac{1}{2\pi}[\arg G]_{\mathcal{L}} = \frac{1}{2i\pi}[\log G]_{\mathcal{L}}, \qquad (5.2.7)$$

which will be called the *index* and does represent the variation of the argument of $G(t)$, as t moves along the contour $\mathcal{L}$ in the positive direction.

Without restricting the generality, we shall suppose that the origin of the coordinate system lies in the domain $\mathcal{L}^+$. Then the function

$$\log[t^{-\chi}G(t)], \quad t \in \mathcal{L},$$

has zero index, is single-valued and satisfies a Hölder condition.

Let

$$\begin{cases} \Gamma(z) &= \dfrac{1}{2i\pi}\displaystyle\int_{\mathcal{L}} \dfrac{\log(t^{-\chi}G(t))dt}{t-z}, \quad z \notin \mathcal{L}, \\[2mm] X^+(z) &= e^{\Gamma(z)}, \quad z \in \mathcal{L}^+, \\ X^-(z) &= z^{-\chi}e^{\Gamma(z)}, \quad z \in \mathcal{L}^-. \end{cases} \qquad (5.2.8)$$

The functions X^+ and X^- are usually referred to as the *canonical* functions of the BVP. Thus $G(t)$ admits the following factorization by (5.2.4)

$$X^+(t) = G(t)X^-(t), \quad t \in \mathcal{L}, \qquad (5.2.9)$$

which gives, *en passant*, a solution of the *homogeneous* BVP (i.e. when $g(t) \equiv 0$) having order $-\chi$ at infinity (i.e. behaving as $z^{-\chi}$). Putting (5.2.9) into (5.2.6), we get

$$\frac{\Phi^+(t)}{X^+(t)} = \frac{\Phi^-(t)}{X^-(t)} + \frac{g(t)}{X^+(t)}, \quad t \in \mathcal{L}. \qquad (5.2.10)$$

Since $\dfrac{g}{X^+} \in \mathbb{H}_\mu(\mathcal{L})$, see [61], we can write, using the Sokhotski-Plemelj formula (5.2.4),

$$\psi^+(t) - \psi^-(t) = \frac{g(t)}{X^+(t)}, \quad t \in \mathcal{L},$$

where

$$\psi(z) = \frac{1}{2i\pi}\int_{\mathcal{L}} \frac{g(s)ds}{X^+(s)(s-z)}, \quad z \notin \mathcal{L}. \qquad (5.2.11)$$

Thus (5.2.10) can be rewritten as

$$\frac{\Phi^+(t)}{X^+(t)} - \psi^+(t) = \frac{\Phi^-(t)}{X^-(t)} - \psi^-(t), \quad t \in \mathcal{L}.$$

Two cases now must be distinguished:

(a) $\chi \geq 0$. From the principle of analytic continuation and Liouville's theorem, we obtain

$$\Phi(z) = X(z)\psi(z) + X(z)P_\chi(z), \quad \forall z \notin \mathcal{L}, \tag{5.2.12}$$

where $X(z)$ and $\psi(z)$ are given by (5.2.8) and (5.2.11), and P_χ is an arbitrary polynomial of degree χ. Thus one sees that, in the case of a closed contour $\mathcal{L}$, the solutions belonging to the class of functions bounded at infinity depend on $\chi + 1$ arbitrary complex constants. The second term in the right member of (5.2.12) does represent the general solution of the homogeneous Riemann BVP.

(b) $\chi < 0$. In this case, the ratio $\dfrac{\Phi^-}{X^-}$ vanishes at infinity, so that $P_\chi \equiv 0$, from Liouville's theorem and

$$\Phi(z) = X(z)\psi(z). \tag{5.2.12bis}$$

But Φ^-, in view of (5.2.8), has a pole of order $-\chi - 1$ at infinity. In order Φ to be holomorphic at infinity (in particular bounded), it is necessary and sufficient that ψ has a zero of order not smaller than $-\chi - 1$.

By expanding the Cauchy-type integral representing ψ (see (5.2.11)) in power series of z around the point at infinity, we get the following conditions of solubility

$$\int_{\mathcal{L}} \frac{g(t)t^{k-1}}{X^+(t)} dt = 0, \quad k = 1, 2, \dots, -\chi - 1. \tag{5.2.13}$$

One can state the following

Theorem 5.2.2   *The number of linearly independent (non trivial) solutions of the homogeneous (i.e. $g = 0$) Riemann BVP (5.2.6) is given by*

$$\ell = \begin{cases} \max(0, \chi + 1), & \chi \neq -1, \\ 1, & \chi = -1, \end{cases} \tag{5.2.14}$$

and the number of conditions of solubility is

$$p = \max(0, -\chi - 1). \tag{5.2.15}$$

∎

- For $\chi = -1$, the non homogeneous problem is always soluble and has a unique solution.

- For $\chi < -1$, the non homogeneous problem has in general no solution. It has exactly one if, and only if, the free term g does satisfy $p = -\chi - 1$ conditions given in (5.2.13).

Let us emphasize that the Riemann BVP (5.2.6) for a closed contour requires in fact to find two functions Φ^+ and Φ^-, which are continuous up the boundary $\mathcal{L}$. Furthermore, it can be shown easily that, if we were to admit solution functions taking infinite values of integrable order on the contour, the class of solutions would in fact not be extended (see [35], p. 41). This assertion is no more true for open contours, as will be shown in the next section, since the behaviour of a Cauchy type integral around the ends of an arc is in general more complicated (see e.g. [35], [61]).

5.2.4 The Riemann BVP for an Open Contour

Here we will quote the material directly used to solve the main functional equation (5.1.1).

Let $\mathcal{L}$ denote a smooth non self-intersecting arc with ends a and b, which can be supposed finite. The positive direction is chosen from a to b. From now on, c will stand for either a or b. The assumptions on G and g and the statement of the BVP are as in section 5.2.3. Since the complex plane cut along $\mathcal{L}$ (the arc $[ab]$) is a simply connected domain, it is important to note that here the BVP is tantamount to finding only one function Φ, holomorphic in the cut plane and having $\mathcal{L}$ as a line of discontinuity, when in the case of a closed contour we had to determine two different holomorphic functions Φ^+ and Φ^-.

The function Φ will be sought in the class of functions holomorphic at infinity, continuous on $\mathcal{L}$ from the left and from the right, with the possible exceptions of the end c, where locally the estimate

$$|\Phi^\pm(t)| < \frac{A}{|t - c|^\alpha}, \quad 0 \le \alpha < 1, \tag{5.2.16}$$

is to hold. Then the Cauchy-type integral

$$\Omega(z) = \frac{1}{2i\pi} \int_{\mathcal{L}} \frac{\varphi(t)dt}{t - z}, \quad z \in \mathbb{C},$$

with $\varphi \in \mathbb{H}_\mu(\mathcal{L})$, has limiting values when z approaches $\mathcal{L}$ from the left or from the right, given from the formulae (5.2.4) and (5.2.5), for all points not coinciding with an end c at which $\varphi(c) \neq 0$.

It can be shown, see e.g. [61] that Ω admits the following representation, in the vicinities of the ends

$$\Omega(z) = \frac{\varepsilon_c \varphi(c)}{2i\pi} \log(z - c) + \Omega_c(z), \tag{5.2.17}$$

where

$$\varepsilon_c = \begin{cases} -1, & \text{if } c \equiv a, \\ +1, & \text{if } c \equiv b, \end{cases}$$

and $\Omega_c(z)$ is a function holomorphic in a neighbourhood of c and continuous at c.

Returning to the Riemann BVP, let

$$\Gamma(z) = \frac{1}{2i\pi} \int_{\mathcal{L}} \frac{\log(G(t))dt}{t-z},$$

(5.2.18)

where $\log G(t)$ stands for any value of this multi-valued function, which varies continuously over $\mathcal{L}$. Let

$$\begin{cases} G(a) & = \rho_a e^{i\theta_a}, \\ \Delta & = [\arg G]_{\mathcal{L}}, \\ \log G(t) & = \log |G(t)| + i \arg G(t), \end{cases}$$

so that

$$\arg G(a) = \theta_a \quad \text{and} \quad \arg G(b) = \theta_a + \Delta.$$

(5.2.19)

According to (5.2.17), we have at each end c, where $c = a$ or b,

$$\Gamma(z) = \varepsilon_c \frac{\log G(c)}{2i\pi} \log(z - c) + \Gamma_c(z),$$

where $\log G(c)$ is given from (5.2.19). Hence

$$e^{\Gamma(z)} = (z-c)^{\lambda_c} e^{\Omega_c(z)},$$

(5.2.20)

with

$$\lambda_a = \frac{-\theta_a}{2\pi} + \frac{i \log \rho_a}{2\pi},$$

(5.2.21)

$$\lambda_b = \frac{\theta_a + \Delta}{2\pi} - \frac{i \log \rho_b}{2\pi}.$$

(5.2.22)

Coming back to the homogeneous BVP (5.2.6) and taking into account condition (5.2.16), one can state the following result:

$$\begin{cases} -2\pi & < \theta_a \leq 0 & \Rightarrow e^{\Gamma(z)} & \text{is bounded at } a; \\ 0 & < \theta_a < 2\pi & \Rightarrow e^{\Gamma(z)} & \text{is unbounded at } a. \end{cases}$$

Assume we are looking for a solution bounded at a. This will be the situation encountered in chapters 5 and 6, in particular when $G(a)$ is a real positive quantity, in which case one shall speak of *automatic boundedness at a*. Let

$$\chi \overset{\text{def}}{=} \left[\frac{\theta_a + \Delta}{2\pi} \right].$$

(5.2.23)

Then the function

$$X(z) = (z - b)^{-\chi}e^{\Gamma(z)} \tag{5.2.24}$$

is bounded at b. The *index* of the homogeneous BVP (5.2.6) is, by definition, the quantity χ given in 5.2.23.

If solutions with integrable singularities around the point b were to be admitted, then the index would be $\chi + 1$, with the corresponding solution of the homogeneous BVP

$$X(z) = (z - b)^{-\chi-1}e^{\Gamma(z)}, \tag{5.2.25}$$

where $\Gamma(z)$ is given by (5.2.18).

Proceeding as in the case of a closed contour, the following result can be established, based on the fact that X, given in (5.2.22), represents a particular solution of the homogeneous BVP, which has order $-\chi$ at infinity, so that

In all cases, $\lim_{z \to \infty} z^\chi X(z) = 1$. The following general result holds.

Theorem 5.2.3   *The general solution of the Riemann BVP (5.2.6), set on a smooth open arc without self-intersection, (in the class of functions bounded at infinity and having possibly singularities of integrable order at the ends), is given by formulae (5.2.12) or (5.2.12bis) and theorem 5.2.2. Here X is taken from (5.2.24) or (5.2.25) and the index χ is defined in (5.2.23).*
Assuming X given by (5.2.24), then (5.2.12) or (5.2.12bis) hold, where ψ has the integral form (5.2.11) in which

$$X^+(t) = (t - b)^{-\chi}e^{\Gamma^+(t)}, \tag{5.2.26}$$

with, from the Sokhotski-Plemelj formulae,

$$\Gamma^+(t) = \frac{1}{2}\log G(t) + \frac{1}{2i\pi}\int_{\mathcal{L}} \frac{\log(G(s))ds}{s - t}, \quad t \in \mathcal{L}. \tag{5.2.27}$$

∎

5.2.5 The Riemann-Carleman Problem with a Shift

We shall need the solution of a generalized BVP, sometimes referred to as the Carleman problem, see [35] [44], hereafter presented under the name *Riemann-Carleman problem*.

Let $\mathcal{L}$ be a simple smooth closed contour. The Riemann-Carleman BVP has the following formulation:

Find a function Φ^+ holomorphic in $\mathcal{L}^+$, the limiting values of which on the contour are continuous and satisfy the relation

$$\boxed{\Phi^+(\alpha(t)) = G(t)\Phi^+(t) + g(t), \quad t \in \mathcal{L},} \tag{5.2.28}$$

where

(i) $g, G \in \mathbb{H}_\mu(\mathcal{L})$, $G(t) \neq 0, \quad \forall t \in \mathcal{L}$;

(ii) $\alpha(t)$, referred to as the *shift* in the sequel, is a function establishing a one to one mapping of the contour $\mathcal{L}$ onto itself, such that the direction of traversing $\mathcal{L}$ is changed and

$$\alpha'(t) = \frac{d\alpha(t)}{dt} \in \mathbb{H}_\mu(\mathcal{L}), \quad \alpha'(t) \neq 0, \quad \forall t \in \mathcal{L}.$$

Most of the time, we shall encounter the so-called *Carleman's condition*

$$\alpha(\alpha(t)) = t, \quad \forall t \in \mathcal{L}. \tag{5.2.29}$$

Whenever needed, this condition will be explicitly mentioned. It is worth remarking now that our situation corresponds to $\alpha(t) = \bar{t}, \forall t \in \mathcal{L}$ (see equation (5.1.3)).

The BVP (5.2.28) amounts in fact to (5.2.6) and allows thus to get an integral representation for the solution functions. But the complexity is now increased, because the density of the Cauchy-type integral solution satisfies a Fredholm integral equation of second kind. We follow in this section the method given in Litvinchuk [44]. The main ideas are presented in the next three theorems.

First, one can easily show that, when (5.2.29) holds, a necessary condition of solvability of (5.2.28) is constituted by one of the two following conditions:

1.

$$\partial(t) = \frac{g(\alpha(t)) + g(t)G(\alpha(t))}{1 - G(t)G(\alpha(t))}, \quad G(t)G(\alpha(t)) \neq 1, \quad \forall t \in \mathcal{L},$$

is the boundary value of a function holomorphic in $\mathcal{L}^+$, in which case one has simply

$$\Phi^+(z) = \frac{1}{2i\pi} \int \frac{\partial(t)dt}{t-z}, \quad z \in \mathcal{L}^+.$$

2.

$$G(t)G(\alpha(t)) = 1 \quad \text{and} \quad g(t) + g(\alpha(t))G(t) = 0. \tag{5.2.30}$$

The latter condition (5.2.30) corresponds in fact to a much richer situation, which we shall encounter. Hence, (5.2.30) will be assumed to hold throughout the sequel of this BVP overview.

Lemma 5.2.4   *The homogeneous Riemann-Carleman BVP*

$$\Phi_1^+(\alpha(t)) - \Phi_2^+(t) = 0, \quad t \in \mathcal{L}, \tag{5.2.31}$$

has no solution, but arbitrary constants.   ■

Proof.   From the Sokhotski Plemelj formulae (5.2.3), the conditions for the analyticity of Φ^+ take the form

$$\frac{1}{2}\Phi_1^+(t) - \frac{1}{2i\pi}\int_{\mathcal{L}}\frac{\Phi_1^+(s)ds}{t-s} = 0, \quad t \in \mathcal{L}. \tag{5.2.32}$$

Substituting t by $\alpha(t)$ aand s by $\alpha(s)$ in (5.2.32), we have

$$\frac{1}{2}\Phi_1^+(\alpha(t)) + \frac{1}{2i\pi}\int_{\mathcal{L}}\frac{\Phi_1^+(\alpha(s))\alpha'(s)ds}{\alpha(s)-\alpha(t)} = 0,$$

or, using (5.2.31),

$$\frac{1}{2}\Phi_2^+(t) + \frac{1}{2i\pi}\int_{\mathcal{L}}\frac{\Phi_2^+(s)\alpha'(s)ds}{\alpha(s)-\alpha(t)} = 0. \tag{5.2.33}$$

Replacing Φ_1 by Φ_2 in (5.2.31) and summing with (5.2.33), we get precisely

$$\Phi_2^+(t) + \frac{1}{2i\pi}\int\left[\frac{\alpha'(s)}{\alpha(s)-\alpha(t)} - \frac{1}{s-t}\right]\Phi_2^+(s)ds = 0, \quad t \in \mathcal{L}. \tag{5.2.34}$$

As can be easily checked, the kernel of (5.2.34) has for $t = s$ a singularity of order less than one, which is consequently integrable, so that the integral operator is compact. It follows that (5.2.34) is a Fredholm integral equation of second kind which thus admits only a finite number of linearly independent solutions.

Assume that (Φ_1, Φ_2) is a pair of non trivial solutions of (5.2.31). Then (Φ_1^k, Φ_2^k) form also a solution for any arbitrary positive integer k. But then Φ_2^k does also satisfy the integral equation (5.2.34), which, since k is arbitrary, would admit an infinite number of independent solutions, contradicting the Fredholm alternative. Thus (5.2.31) admits only the solutions $\Phi_1 = \Phi =$ Constant. Lemma 5.2.4 is proved.   ∎

Let us introduce the operator $\mathcal{B}$,

$$(\mathcal{B}\varphi)(t) \equiv \varphi(t) + \frac{1}{2i\pi}\int_{\mathcal{L}}\left[\frac{1}{s-t} - \frac{\alpha'(s)}{\alpha(s)-\alpha(t)}\right]\varphi(s)ds. \tag{5.2.35}$$

Then the following lemma holds.

Lemma 5.2.5   *The Fredholm integral equation*

$$(\mathcal{B}\varphi)(t) = 0 \tag{5.2.36}$$

admits only the trivial solution $\varphi \equiv 0$.   ∎

Proof.   Let $\varphi(t)$ be a solution of (5.2.36) and set $\beta(t) = \alpha^{-1}(t)$, the inverse mapping. We follows [44]. Consider the two functions

$$\begin{cases} \Phi_1^+(z) &= \dfrac{1}{2i\pi} \displaystyle\int_{\mathcal{L}} \dfrac{\varphi(\beta(s))ds}{s-z}, \quad z \in \mathcal{L}^+, \\[4mm] \Phi_2^+(z) &= \dfrac{-1}{2i\pi} \displaystyle\int_{\mathcal{L}} \dfrac{\varphi(s)ds}{s-z}, \quad z \in \mathcal{L}^+. \end{cases}$$

It follows from (5.2.4) and (5.2.5) that

$$(\mathcal{B}\varphi)(t) = \Phi_1^+(\alpha(t)) - \Phi_2^+(t) = 0, \quad t \in \mathcal{L},$$

whence, using lemma 5.2.4,

$$\Phi_1^+ = \Phi_2^+ = C,$$

where C is an arbitrary constant. Thus

$$\frac{1}{2i\pi} \int_{\mathcal{L}} \frac{\varphi(\beta(s)) - C}{s-z}ds = \frac{-1}{2i\pi} \int_{\mathcal{L}} \frac{\varphi(s) + C}{s-z}ds = 0, \quad z \in \mathcal{L}^+,$$

so that, by Cauchy's theorem,

$$\begin{aligned} \varphi(\beta(t)) - C &= \psi_1^-(t), \quad t \in \mathcal{L}, \\ \varphi(t) + C &= \psi_2^-(t), \quad t \in \mathcal{L}, \end{aligned}$$

where $\psi_i^-(t)$ is holomorphic in $\mathcal{L}^-$, with the condition $\psi_i^-(\infty) = 0$, $i = 1, 2$. Moreover, we have

$$W_1^-(\alpha(t)) = W_2^-(t), \quad t \in \mathcal{L},$$

with

$$\begin{cases} W_1^- = \psi_1^- + C, \\ W_2^- = \psi_2^- - C. \end{cases}$$

Clearly it can be shown as in lemma 5.2.4, that the *exterior* problem (5.2.31) (in $\mathcal{L}^-$) has only trivial solutions

$$W_1^-(z) = W_2^-(z) = C_1.$$

Hence, using the conditions $\psi_1(\infty) = \psi_2(\infty) = 0$, we obtain $C = C_1 = -C$. Consequently, $C_1 = C = 0 = \psi_i(z) = \varphi(t)$. Lemma 5.2.5 is proved. ∎

Sometimes it happens that, for special relationships between the parameter values of the random walk (see for example [26, 27]), one has to solve problem (5.2.28) for $G(t) \equiv 1$, which belongs then to the so-called *Dirichlet-Carleman* class described thereafter.

Theorem 5.2.6   *The Carleman-Dirichlet problem*

$$\Phi^+(\alpha(t)) - \Phi^+(t) = g(t), \quad t \in \mathcal{L}, \tag{5.2.37}$$

where $g \in \mathbb{H}_\mu(\mathcal{L})$ satisfies the relation

$$g(t) + g(\alpha(t)) = 0,$$

has a unique solution given up to an arbitrary additive constant by

$$\Phi^+(z) = \frac{1}{2i\pi} \int_{\mathcal{L}} \frac{\varphi(\alpha(s))ds}{s - z} + C, \qquad (5.2.38)$$

where C is an arbitrary constant and $\varphi(t)$ is the unique solution of the integral equation

$$(\mathcal{B}\varphi)(t) = g(t),$$

$\mathcal{B}$ *being the operator defined in (5.2.35).*

Proof.   It is a direct consequence of lemmas 5.2.4 and 5.2.5 and of the Fredholm alternative.   ∎

We return now to the general problem (5.2.28), which corresponds to the main equation (5.1.3). There are several ways to produce a solution having an integral representation. The most direct one (which we will choose) is proposed in [44], and relies on the following theorem of intrinsic interest about *conformal gluing.*

Theorem 5.2.7   *Let $\alpha(t)$ be a Carleman automorphism of the curve $\mathcal{L}$. Then there exists a function w, holomorphic in $\mathcal{L}^+$, except at one point $z = z_0 \in \mathcal{L}^+$, where w has a simple pole, such that*

$$w(\alpha(t)) - w(t) = 0, \quad t \in \mathcal{L}. \qquad (5.2.39)$$

Moreover w establishes a conformal mapping of the domain $\mathcal{L}^+$ onto the domain Δ, which consists of the plane cut along an open smooth arc $\mathcal{U}$.

Proof.   The proof of this theorem can be found in [44].   ∎

From theorem 5.2.7, one can write

$$w(z) = \frac{1}{z - z_0} + \psi^+(z),$$

where $\psi^+(z)$ is holomorphic in $\mathcal{L}^+$ and satisfies the condition

$$\psi^+(\alpha(t)) - \psi^+(t) = \frac{1}{t - z_0} - \frac{1}{\alpha(t) - z_0} \overset{def}{\equiv} g(t), \quad t \in \mathcal{L}.$$

Here (5.2.30) is fulfilled, so that ψ^+ is given by the formula (5.2.38).
It can be shown see [44] that α has two fixed points. In the situation we shall encounter, they are simply the intersection points of $\mathcal{L}$ with the real axis (by symmetry arguments) denoted by A and B as in figure 5.2.1.
We denote by $\mathcal{L}_u$ (resp. $\mathcal{L}_d$) the upper part (resp. lower) of $\mathcal{L}$ described in positive direction, i.e. the one which, on traversing the boundary, leaves the domain $\mathcal{L}^+$ on the left. The respective ends of $\mathcal{U}$ are denoted by a and b.

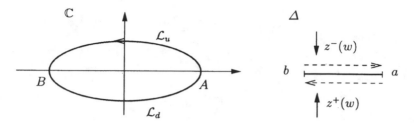

Fig. 5.2.1.

From theorem 5.2.7, it follows that the function w, see (5.2.39), has an inverse denoted by z, satisfying

$$\begin{cases} z^+(w) &= \alpha(t) \\ z^-(w) &= t \end{cases} , \quad \text{for } \omega \in \mathcal{U} \text{ and } t \in \mathcal{L}_d.$$

On $\mathcal{L}_u$, one must exchange z^+ and z^- and $\mathcal{L} \overset{w(z)}{\underset{z(w)}{\rightleftharpoons}} \mathcal{U}$.

Now the boundary condition (5.2.28) can be rewritten as

$$\Phi^+[z^+(w)] = G[z^-(w)]\Phi^+[z^-(w)] + g(z^-(w)), \quad w \in \mathcal{U}. \tag{5.2.40}$$

Introduce the function $\theta = \Phi^+ \circ z$,

$$\theta(w) \overset{\text{def}}{=} \Phi^+[z(w)], \quad w \in \mathbb{C},$$

the limiting values of which, for w approaching $\mathcal{U}$, satisfy

$$\begin{cases} \theta^+(w) = \Phi^+[z^+(w)], \\ \theta^-(w) = \Phi^+[z^-(w)]. \end{cases}$$

Then the BVP (5.2.28) on the closed contour $\mathcal{L}$ has been reduced to a Riemann BVP on the arc $\mathcal{U}$ (analyzed in section 5.2.4), with the boundary condition

$$\theta^+(w) = G[z^-(w)]\theta^-(w) + g[z^-(w)], \quad w \in \mathcal{U}. \tag{5.2.41}$$

Keeping in mind that we are interested by solutions of (5.2.40) which are *bounded at both ends*, we shall compute the index of the problem solely in this case.

When w describes $\mathcal{U}$ from a to b in the positive direction, as shown on figure 5.2.1, the point $z^-(w)$ describes the arc $\mathcal{L}_d$ in the negative direction. Thus

$$[\arg G \circ z^-]_\mathcal{U} = -[\arg G]_{\mathcal{L}_d}.$$

On the other hand,

$$0 = [\arg G]_{\mathcal{L}_d} + [\arg G \circ \alpha]_{\mathcal{L}_d} = [\arg G]_{\mathcal{L}_d} - [\arg G]_{\mathcal{L}_u},$$

so that

$$[\arg G]_{\mathcal{L}_d} = [\arg G]_{\mathcal{L}_u}.$$

Setting, as in (5.2.7),

$$\chi = \frac{1}{2\pi}[\arg G]_{\mathcal{L}},$$

we have

$$\begin{aligned}[\arg G]_{\mathcal{L}} &= [\arg G]_{\mathcal{L}_d} + [\arg G]_{\mathcal{L}_u}\\ &= 2[\arg G]_{\mathcal{L}_d},\end{aligned}$$

whence

$$\frac{1}{2\pi}[\arg G \circ z^-]_u = \frac{-\chi}{2}. \tag{5.2.42}$$

Note that χ is even and, in the situation considered above,

$$G(t_A) = G(t_B) = 1,$$

where t_A and t_B are the affixes of A and B respectively. Moreover

$$[\arg G]_{\mathcal{L}_d} = \arg G(t_B) - \arg G(t_A) = 2k\pi.$$

On the other hand, the index $\tilde{\chi}$ of the problem (5.2.41) has been given in (5.2.23)

$$\tilde{\chi} = \frac{\theta_a + \Delta}{2\pi} = \frac{\Delta}{2\pi} \quad (\text{since } \theta_a = 0),$$

with

$$\Delta = [\arg G(z^-(w))]_u = -\chi\pi.$$

Thus

$$\tilde{\chi} = \frac{-\chi}{2} \tag{5.2.43}$$

and we can state the final following result.

Theorem 5.2.8

(i) *If $\tilde{\chi} \geq 0$, then the BVP (5.2.28) has $\ell = 1 + \tilde{\chi}$ linearly independent solutions bounded at both ends.*

(ii) *If $\tilde{\chi} < 0$, then the homogeneous problem (5.2.28) has no solution and the non homogeneous problem is soluble if, and only if, $p = -\tilde{\chi} - 1$ conditions of the form (5.2.13) are fulfilled. In this case, using (5.2.12bis), the solution of the BVP (5.2.28), which has been reduced to the BVP (5.2.41), is given by*

$$\begin{aligned}\Phi^+(z) \equiv \theta(w(z)) &= \frac{X(w(z))}{2i\pi}\int_{[ab]}\frac{g(z^-(s))ds}{X^+(s)(s-w(z))} \tag{5.2.44}\\ &= \frac{-X(w(z))}{2i\pi}\int_{\mathcal{L}_d}\frac{w'(t)g(t)dt}{X^+(w(t))(w(t)-w(z))}, \forall z \in \mathcal{L}^+.\end{aligned}$$

■

5.3 Further Properties of the Branches Defined by $Q(x, y) = 0$

With the notation of section 2.3 of chapter 2, let us recall that

$$Q(x, y) = xy[P(x, y) - 1] \equiv a(x)y^2 + b(x)y + c(x) \equiv \tilde{a}(y)x^2 + \tilde{b}(y)x + \tilde{c}(y).$$

Here we shall analyze more completely the branches $Y_0(x)$ and $Y_1(x)$ of the algebraic function $Y(x)$.

It will be assumed, unless otherwise mentioned, that $Q(x, y)$ is irreducible of degree 2 with respect to x and y, that the Riemann surface has genus 1 and also that there are no branch points on the unit circle Γ. The case of genus 0 will be the subject of chapter 6.

Lemma 5.3.1   *The equation $Q(x, y) = 0$ has two roots $Y_0(x)$ and $Y_1(x)$, such that, for $|x| = 1$,*

$$\begin{cases} |Y_0(x)| \leq 1, \\ |Y_1(x)| \geq 1, \end{cases} \tag{5.3.1}$$

with strict inequalities, except for $x = 1$, where we have

$$\begin{cases} Y_0(1) = \min\left(1, \dfrac{\sum_i p_{i,-1}}{\sum_i p_{i,1}}\right), \\ Y_1(1) = \max\left(1, \dfrac{\sum_i p_{i,-1}}{\sum_i p_{i,1}}\right) \end{cases} \tag{5.3.2}$$

∎

Proof.   Let $|x| = 1$, $x \neq 1$. Then $\Re[-1 + P(x, y)] < 0$, $\forall y$ such that $|y| = 1$, where $\Re[z]$ denotes the real part of the complex number z, whence

$$[\arg(-1 + P(x, .))]_\Gamma = 0, \tag{5.3.3}$$

where Γ is the unit circle in the complex plane $\mathbb{C}_y$. From the main assumptions made at the beginning of this section, $\sum_i p_{i1} > 0$. Since $P(x, y)$ has a simple pole at $y = 0$, the principle of the argument in (5.3.3) shows that the equation $Q(x, y) = 0$, $|x| = 1$, $x \neq 1$ has exactly one root $y = Y_0(x)$ inside the unit disk $\mathcal{D}$. For $x = 1$, one distorts Γ by making a small indentation to the left [resp. right] of the point $y = 1$ inside [resp. outside] Γ, when $M_y > 0$ [resp. $M_y < 0$], so that (5.3.3) still holds on this new contour, say $\tilde{\Gamma}$. This proves (5.3.2) and lemma 5.3.1. ∎

As an immediate consequence, we have the following:

Corollary 5.3.2

(a) When $\sum_i p_{i1} = 0$, $Q(x, y)$ is of first degree in y and the root $Y_0(x)$ satisfies
(5.3.1) with $Y_0(1) = 1$.

(b) When $\sum_i p_{i,-1} = 0$, then $c(x) = 0$ and

$$Q(x, y) = a(x)y^2 + b(x)y = y[a(x) + b(x)].$$

The algebraic curve would then be decomposed, with one trivial root $Y_0(x) = 0$,
$\forall x$. The other root Y_1 does satisfy (5.3.1) and $Y_1(1) = 1$.

In the next lemmas, specific properties of the functions $Y_0(x)$ and $Y_1(x)$ will
be given. In particular, one will study the transforms $Y_i[x_1 x_2]$ and $Y_i[x_3 x_4]$,
$i = 1, 2$, of the real slits $[x_1 x_2]$ and $[x_3 x_4]$, which will be shown to be two simple
non intersecting closed curves in the complex plan $\mathbb{C}_y$.

Notation

1. From now on, $\mathbb{C}_x$ [resp. $\mathbb{C}_y$], cut along $[x_1 x_2] \cup [x_3 x_4]$ [resp. $[y_1 y_2] \cup [y_3 y_4]$]
   will be denoted by $\widetilde{\mathbb{C}}_x$ [resp. $\widetilde{\mathbb{C}}_y$]. Also, quite naturally, we shall write
   $Y(x)$, whenever $x = x_i, i = 1, \dots, 4$.

2. Occasionally, the following convention will be *ad libitum* employed: $\overrightarrow{x_1 x_2}$
   will stand for the *contour* $[x_1 x_2]$, traversed from x_1 to x_2 along the upper
   edge of the slit $[x_1 x_2]$ and then back to x_1, along the lower edge of the
   slit. Similarly, $\overleftarrow{x_1 x_2}$ is defined by exchanging "upper" and "lower".

3.
$$\begin{cases} \mathcal{L} & = Y_0[\overrightarrow{x_1 x_2}] = \overline{Y}_1[\overleftarrow{x_1 x_2}], \\ \mathcal{L}_{ext} & = Y_0[\overrightarrow{x_3 x_4}] = \overline{Y}_1[\overleftarrow{x_3 x_4}], \\ \mathcal{M} & = X_0[\overrightarrow{y_1 y_2}] = \overline{X}_1[\overleftarrow{y_1 y_2}], \\ \mathcal{M}_{ext} & = X_0[\overrightarrow{y_3 y_4}] = \overline{X}_1[\overleftarrow{y_3 y_4}]. \end{cases}$$

4. For any arbitrary simple closed curve $\mathcal{U}$, $G_{\mathcal{U}}$ [resp. $G_{\mathcal{U}}^c$] will denote the
   interior [resp. exterior] domain bounded by $\mathcal{U}$, i.e. the domain remaining
   on the lefthand side when $\mathcal{U}$ is traversed in the positive (counter clockwise)
   direction. This definition remains valid for the case when $\mathcal{U}$ is unbounded
   but closed at infinity.

Theorem 5.3.3   *The following topological and algebraic properties hold.*

(i) *The curves $\mathcal{L}$ and $\mathcal{L}_{ext}$ (resp. $\mathcal{M}$ and $\mathcal{M}_{ext}$) are simple, closed and symmetrical about the real axis in the $\mathbb{C}_y$ [resp. $\mathbb{C}_x$] plane. They do not intersect if the group of the random walk is not of order 4. When this group is of order 4, $\mathcal{L}$ and $\mathcal{L}_{ext}$ [resp. $\mathcal{M}$ and $\mathcal{M}_{ext}$] coincide and form a circle possibly degenerating into a straight line. In the general case they build the two components (possibly identical and then the circle must be counted twice) of a quartic curve. Moreover, setting*

$$\Delta = \begin{vmatrix} p_{11} & p_{10} & p_{1,-1} \\ p_{01} & p_{0,-1} & p_{0,-1} \\ p_{-1,1} & p_{-1,0} & p_{-1,-1} \end{vmatrix},$$

we have the following topological invariants:
– If $\Delta > 0$, then

$$[y_1 y_2] \subset G_{\mathcal{L}} \subset G_{\mathcal{L}_{ext}} \quad \text{and} \quad [y_3 y_4] \subset G^c_{\mathcal{L}_{ext}};$$

– If $\Delta < 0$, then

$$[y_1 y_2] \subset G_{\mathcal{L}_{ext}} \subset G_{\mathcal{L}} \quad \text{and} \quad [y_3 y_4] \subset G^c_{\mathcal{L}}.$$

– If $\Delta = 0$, then we already know from chapter 4 that $G_{\mathcal{L}}$ is a circular domain, and in fact

$$[y_1 y_2] \subset G_{\mathcal{L}} \equiv G_{\mathcal{L}_{ext}} \quad \text{and} \quad [y_3 y_4] \subset G^c_{\mathcal{L}}.$$

Entirely similar results hold for $\mathcal{M}$, $\mathcal{M}_{ext}$, $[x_1 x_2]$ and $[x_3 x_4]$.

(ii) *The functions Y_i [resp. X_i], $i = 0, 1$, are meromorphic in the cut plane $\mathbb{C}_x$ [resp. $\mathbb{C}_y$]. In addition,*

- *Y_0 [resp. X_0] has two zeros and no poles.*
- *Y_1 [resp. X_1] has two poles and no zeros.*
- *$|Y_0(x)| \leq |Y_1(x)|$ [resp. $|X_0(y)| \leq |X_1(y)|$], in the whole cut complex plane. Equality takes place only on the cuts.*

(iii) *The function Y_0 can become infinite at a point x if, and only if,*

$$p_{11} = p_{10} = 0, \quad \text{and then} \quad x = x_4 = \infty,$$

or

$$p_{-11} = p_{-10} = 0, \quad \text{and then} \quad x = x_1 = 0.$$

In addition, most of the properties cited in (ii) and (iii) hold in the genus zero case. ∎

Proof of (i) .   We have seen in chapter 2 that $Y(x)$ has 4 real branch points, which satisfy either of the following inequalities

$$-1 < x_1 < x_2 < 1 < x_3 < x_4, \tag{5.3.4}$$

$$x_4 < -1 < x_1 < x_2 < 1 < x_3. \tag{5.3.5}$$

It is important to recall that x_2 and x_3 are always positive. When (5.3.4) holds, the $\mathbb{C}_x$ plane is cut along $[x_1 x_2]$ and $[x_3 x_4]$. On the other hand, in the instance (5.3.5), the cut $[x_3 x_4]$ does include the real point at infinity.

As soon as we will have proved that $\mathcal{L} \cup \mathcal{L}_{ext}$ is a quartic, the fact that $\mathcal{L}$ and $\mathcal{L}_{ext}$ are simple curves will become immediate. That they do not intersect in the $\mathbb{C}_y$ plane is less obvious and follows from general properties concerning the existence of double points of quartic curves (see e.g. Hartshorne [37] or Kendig [41]). For the sake of completeness we shall give now a direct analytical demonstration of part (i) for the curves $\mathcal{L}$ and $\mathcal{L}_{ext}$.

Let $y = u + iv$ a point in the complex plane $\mathbb{C}_y$. For $x \in [x_1 x_2] \cup [x_3 x_4]$, $Y_0(x)$ and $Y_1(x)$ are complex conjugate. Thus

$$Y_0(x)Y_1(x) \;=\; \frac{c(x)}{a(x)} = u^2 + v^2,$$

$$Y_0(x) + Y_1(x) \;=\; \frac{-b(x)}{a(x)} = 2u.$$

Hence, we get

$$\begin{cases} (u^2 + v^2)a(x) - c(x) = 0, \\ 2ua(x) + b(x) = 0. \end{cases} \tag{5.3.6}$$

The two polynomial equations of second degree in system (5.3.6) must have a common root. Recall briefly that two equations in t, say

$$\begin{cases} \alpha_1 t^2 + \beta_1 t + \gamma_1 = 0, \\ \alpha_2 t^2 + \beta_2 t + \gamma_2 = 0, \end{cases}$$

have a common root if, and only if, their coefficients satisfy the relation

$$(\alpha_1 \gamma_2 - \alpha_2 \gamma_1)^2 + (\alpha_1 \beta_2 - \alpha_2 \beta_1)(\gamma_1 \beta_2 - \gamma_2 \beta_1) = 0.$$

We apply this relation to (5.3.6), instantiating

$$\begin{cases} \alpha_1 = (u^2 + v^2)p_{11} - p_{1,-1}, \\ \beta_1 = (u^2 + v^2)p_{01} - p_{0,-1}, \\ \gamma_1 = (u^2 + v^2)p_{-1,1} - p_{-1,-1}, \end{cases} \qquad \begin{cases} \alpha_2 = 2up_{11} + p_{1,0}, \\ \beta_2 = 2up_{01} + (P_{00} - 1), \\ \gamma_2 = 2up_{-1,1} + P_{-1,0}. \end{cases}$$

Then the equation of the quartic curve in the $\mathbb{C}_y$ plane, with coordinates (u, v), reads

$$R^2 + ST = 0, \tag{5.3.7}$$

where

$$R = (u^2 + v^2)(p_{11}p_{-1,0} - p_{-1,1}p_{10})$$
$$+ 2u(p_{11}p_{-1,-1} - p_{1,-1}p_{-1,1}) + p_{10}p_{-1,-1} - p_{-1,0}p_{1,-1},$$

$$S = (u^2 + v^2)(p_{11}(p_{00} - 1) - p_{01}p_{10})$$
$$+ 2u(p_{11}p_{0,-1} - p_{1,-1}p_{01}) + p_{10}p_{0,-1} + (1 - p_{00})p_{1-1},$$

$$T = (u^2 + v^2)(p_{-1,1}(p_{00} - 1) - p_{01}p_{-1,0})$$
$$+ 2u(p_{-1,1}p_{0,-1} - p_{-1,-1}) + p_{-1,0}p_{0,-1} + (1 - p_{00})p_{-1,1}.$$

As mentionned earlier it is possible, from (5.3.7), to write conditions for this quartic to have no double points. Nevertheless we shall present thereafter a mild force argument proving directly this assertion.

The curves $\mathcal{L}$ and $\mathcal{L}_{ext}$ are symmetrical about the real axis. Assume for a while that they have an intersection or a common vertical tangent at a point of the real axis. Then system (5.3.6) holds for two distinct values of x, say x' and x'', but for one and the same pair $(u, |v|)$. Furthermore, one can always find a triple of complex numbers $(\alpha, \beta, \gamma) \neq (0, 0, 0)$, such that the polynomial

$$F(x) = \alpha a(x) + \beta b(x) + \gamma c(x)$$

has zeros at x' and at an arbitrary fixed point $x_0 \neq x', x''$. But then $F(x)$ also has a zero at x'', since $\dfrac{b(x')}{a(x')} = \dfrac{b(x'')}{a(x'')}$ and $\dfrac{c(x')}{a(x')} = \dfrac{c(x'')}{a(x'')}$, using (5.3.6)

But then, the equation $F(x) = 0$ would have three distinct roots, which is impossible, since $F(x)$ is a polynomial of second degree in x, unless $F(x) \equiv 0$, which means that $a(x)$, $b(x)$ and $c(x)$ are linearly dependant. Consequently the group of the random walk is of order 4, as shown in lemma 4.4.2 of chapter 4 and, in this case, the curves $\mathcal{L}$ and $\mathcal{L}_{ext}$ coincide to form a circle (the quartic is then a circle counted twice).

To prove the last point of *(i)* and *(ii)* in theorem 5.3.3, it is necessary to know the sign of the quantities $Y(x_i)$, $i = 1, \dots, 4$. In $\widetilde{\mathbb{C}}_x$, $Y_i(x)$, $i = 0, 1$, are meromorphic. Since, by (5.3.4), (5.3.5), there are no branch points on the real interval $]x_2, x_3[$ and $Y_0(1) < Y_1(1)$, we have

$$Y_0(x) < Y_1(x), \quad \forall x \in]x_2, x_3[, \tag{5.3.8}$$

remembering that Y_0 and Y_1 are real on $]x_2, x_3[$. Moreover, Y_0 and Y_1 have no zeros and no poles on $]x_2, x_3[$, since otherwise either $a(x)$ or $c(x)$ would vanish for positive values of x, which is impossible. This argument shows that Y_i, $i = 0, 1$, does not vanish on $\mathbb{R}^+ - \{0 \cup \infty\}$. Thus the quantities $Y_i(x_2)$ and $Y_i(x_3)$ have the same sign, which is the one of $Y_i(1)$, i.e. positive, and they are finite.

• *Sign of* $Y(x_1)$ First, we observe that $b(x) = p_{10}x^2 + (p_{00} - 1)x + p_{-10}$ has two real positive roots (one being rejected at infinity for $p_{10} = 0$). This follows from

$$b(0) \geq 0, \quad b(1) < 0, \quad b(+\infty) = +\infty.$$

Moreover, since $b(x) = 0 \Rightarrow b^2(x) - 4a(x)c(x) < 0$, one root belongs to the interval $]x_1 x_2[$ and the second one is on $]x_3 x_4[$, noting that $x_4 = \infty$ if $p_{10} = 0$. Thus $Y(x_1) = \dfrac{-b(x_1)}{2a(x_1)} \leq 0$. The borderline case $Y(x_1) = 0$ is obtained when x_1 is at one and the same time a root of $b(x)$ and of $c(x)$. This implies $x_1 = 0$ and $p_{-1,0} = p_{-1,-1} = 0$.

An other special situation arises when $a(x_1) = b(x_1) = 0$, yielding necessarily $x_1 = 0$ and also $p_{-11} = p_{-10} = 0$. Then $Y(x_1) = \infty$ and we have

$$\Re(Y(0)) = \frac{1 - p_{00}}{2p_{01}} > 0,$$

together with

$$\Im(Y(x)) \sim \frac{\pm 1}{\sqrt{x}} \sqrt{\frac{p_{-1,-1}}{p_{01}}},$$

when $x \to 0_+$, and the curve $\mathcal{L}$ is unbounded.

• *Sign of* $Y(x_4)$ The arguments are quite similar. Now we have

$$Y(x_4) = \frac{-b(x_4)}{a(x_4)}.$$

− If $1 < x_4 < \infty$, then we get $Y(x_4) < 0$, since $b(x_4) > 0$ and $a(x_4) > 0$, using the fact that the possible real zeros of b are necessarily located on the cuts.

− If $-\infty < x_4 < -1$, then the point at infinity on the real line belongs to the cut $[x_3 x_4]$ (using the convention made at the beginning of this section), so that

$$Y_0(\infty) = \overline{Y_1(\infty)}.$$

Thus, provided that $p_{10} + p_{11} \neq 0$,

$$\Re(Y(\infty)) = \frac{-b(\infty)}{a(\infty)} \leq 0,$$

which yields again

$$\Re(Y(x_4)) = Y(x_4) < 0.$$

The limiting case $x_4 = \infty$ implies necessarily $y_4 < \infty$, together with

$$p_{10} = p_{11} = 0 \quad \text{or} \quad p_{10} = p_{1,-1} = 0,$$

which we analyze separately.

- $p_{10} = p_{11} = 0$. Then $x_4 = \infty$ is a common root of $a(x) = b(x) = 0$ and $Y(x_4) = \infty$, which says that the curve $\mathcal{L}_{ext}$ is unbounded. More precisely

$$\begin{cases} \Re(Y(x_4)) = \dfrac{1 - p_{00}}{2p_{01}} \quad \text{giving a vertical asymptote of } \mathcal{L}_{ext}, \\[3mm] \Im(Y(x)) \sim \pm\sqrt{\dfrac{xp_{1,-1}}{p_{01}}}, \quad \text{as } x \to x_4 = \infty. \end{cases}$$

- $p_{10} = p_{1,-1} = 0$. Then $x_4 = \infty$ is a common root of $b(x) = c(x) = 0$ and in this case $Y(x_4) = 0$. The behaviour of $Y(x)$ can be obtained exactly as above, and details will be left aside.

It might be useful to recall that we omit all *singular* random walks introduced in section 2.3, which occur e.g. when

$$p_{11} = p_{10} = p_{1,-1} = 0.$$

Note also the amusing case

$$p_{11} = p_{10} = p_{-10} = p_{-11},$$

yielding a group of order 4, for which $\mathcal{L} \equiv \mathcal{L}_{ext}$ is simply the vertical line of abscissa $\dfrac{1 - p_{00}}{2p_{01}}$.

Now it is the right moment to deal with statement *(iii)* of the theorem, since it overlaps what has been done just above.

Proof of (iii).   Assume there exists x with $|Y_0(x)| = \infty$. Then, anticipating slightly the results of *(ii)*, it follows that $|Y_1(x)| = \infty$. Since

$$Y_0(x) + Y_1(x) = \frac{-b(x)}{a(x)} \quad \text{and} \quad Y_0(x)Y_1(x) = \frac{c(x)}{a(x)},$$

we must have $a(x) = b(x) = 0$, so that either $x = x_1$ or $x = x_4$, because the roots of a (resp. b) cannot be positive (resp. negative).

- $a(x_1) = b(x_1) = 0$. Then $p_{-11} = p_{-10} = 0$ and the curve $\mathcal{L}$ is unbounded.

- $a(x_4) = b(x_4) = 0$. Then $p_{11} = p_{10} = 0$. This case was encountered above, implying $x_4 = \infty$ and the curve $\mathcal{L}_{ext}$ unbounded.

This result will be used in chapter 6 for the genus 0 case.

We continue with the last part of assertion *(i)* in the theorem, by using arguments similar to those of section 2.3, namely the continuity with respect to the parameters p_{ij}. First let us denote by $\mathcal{A}$ the set of points in the simplex $\mathcal{P}$ for which $M_x = M_y = 0$. This set $\mathcal{A}$ does subdivide $\wp$ into 4 convex sets $\mathcal{A}_{ij}$, for $i, j = 0, 1$. Now consider the hypersurface $\mathcal{H}$

$$\Delta = \begin{vmatrix} p_{11} & p_{10} & p_{1,-1} \\ p_{01} & p_{00} - 1 & p_{0,-1} \\ p_{-1,1} & p_{-1,0} & p_{-1,-1} \end{vmatrix} = 0.$$

$\mathcal{H}$ divides each $\mathcal{A}_{ij}$ into two pathwise connected subsets $\mathcal{A}_{ij}^+$ and $\mathcal{A}_{ij}^-$, for which $\Delta > 0$ and $\Delta < 0$, respectively.

It suffices to check now 3 special cases for which, respectively, $\Delta = 0$, $\Delta > 0$ and $\Delta < 0$:

(a) The usually called *simple* random walk. Here we know that $\Delta = 0$. Moreover $\mathcal{L} \equiv \mathcal{L}_{ext}$ is the circle centered at the origin, with radius $\sqrt{\frac{p_{0,-1}}{p_{0,1}}}$. It is not difficult to check that the slit $[y_1 y_2]$ (resp. $[y_3 y_4]$) lies always inside (resp. outside) this circle, whenever M_x ad M_y do not simultaneously vanish.

(b)

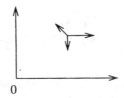

This random walk can depict two $M/M/1$ queues in tandem and its corresponding group is of order 6. Then

$$\Delta = \begin{vmatrix} 0 & p_{10} & 0 \\ 0 & -1 & p_{0-1} \\ p_{-11} & 0 & 0 \end{vmatrix} = p_{-11} p_{10} p_{0-1} > 0.$$

Assuming M_x or M_y different from zero (i.e. genus 1), the curves $\mathcal{L}$, $\mathcal{L}_{ext}$ are shown in figure 5.3.1 $G_{\mathcal{L}} \subset G_{\mathcal{L}_{ext}}$. Here $[y_1 y_2] \subset G_{\mathcal{L}}$ and $[y_3, \infty] \subset G^c_{\mathcal{L}_{ext}}$. Analogous results hold for $\mathcal{M}$ and $\mathcal{M}_{ext}$ in the complex plane $\mathbb{C}_x$.

(c)

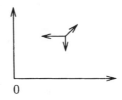

Here $\Delta = \begin{vmatrix} p_{11} & 0 & 0 \\ 0 & -1 & p_{0,-1} \\ 0 & p_{-10} & 0 \end{vmatrix} = -p_{11}p_{0,-1}p_{-10} < 0.$

This random walk was already presented in section 4.7 and has also a group of order 6. Here we have $p_{10} = p_{1-1} = 0$, so that, from the preceding analysis, $x_4 = \infty$ and $Y(x_4) = 0$, assuming again $M_x \neq 0$ or $M_y \neq 0$. As shown in figure 5.3.1, $G_{\mathcal{L}_{ext}} \subset G_{\mathcal{L}}$, $[y_1 y_2] \subset G_{\mathcal{L}_{ext}}$ and $[y_3 y_4] \subset G_{\mathcal{L}}^c$.

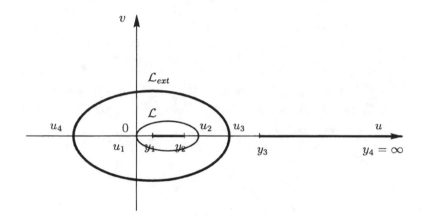

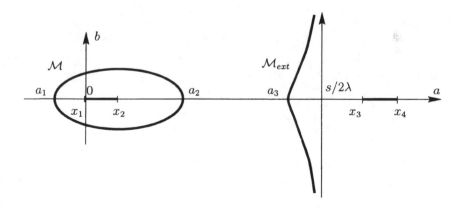

Fig. 5.3.1.

Returning to the general random walk, we note that any point $\rho \in \mathcal{A}_{ij}^+$ can be connected along a continuous path $\ell \subset \mathcal{A}_{ij}^+$ to some point $\rho_0 \in \mathcal{A}_{ij}^+$ corresponding to the particular random walk (b). But the respective positions of the curves and the slits can be derived from the principle of the argument applied to the function

$$\widetilde{D}(y) = \widetilde{b}^2(y) - 4\widetilde{a}(y)\widetilde{c}(y)$$

along the curve $\mathcal{L}_{ext}$. Since $[\arg \widetilde{D}(y)]_{\mathcal{L}_{ext}}$ is a continuous function of the p_{ij}'s, taking discrete values in the region $\mathcal{A}_{ij}^+$, we conclude that, whenever $\Delta > 0$ (in the genus 1 case), $[y_1 y_2] \subset G_{\mathcal{L}} \subset G_{\mathcal{L}_{ext}}$ and $[y_3 y_4] \subset G_{\mathcal{L}_{ext}}^c$.
The case $\Delta < 0$, using the regions $\mathcal{A}_{ij}^-$ and the random walk (c) can be treated in a similar way. This concludes the part (i) of theorem 5.3.3.

Remark 5.3.4 *As shown in section 2.5, the curves $\mathcal{L}$, $\mathcal{L}_{ext}$, $\overrightarrow{y_1 y_2}$ and $\overrightarrow{y_3 y_4}$ belong to the same homotopy class on the Riemann surface* **S**. *This indicates a priori that the curves $\mathcal{L}$ or $\mathcal{L}_{ext}$ contains in their interior at most one slit.*

Proof of (ii)   We have already shown in the proof of (i) that $Y(x_2)$ is positive. Owing to (5.3.8), it follows that $Y_0[\overrightarrow{x_1 x_2}]$ is the curve $\mathcal{L}$ traversed in the positive direction. Since on $[x_1 x_2]$, $Y_1(x)$ is the conjugate of $Y_0(x)$, $Y_1[\overrightarrow{x_1 x_2}]$ is the curve $\mathcal{L}$ traversed in the negative (clockwise) direction. Similar conclusions hold for the curve $\mathcal{L}_{ext}$. It follows that

$$\begin{cases} \mathcal{I}nd[Y_0(x)]_{\{\overrightarrow{x_1 x_2} \cup \overrightarrow{x_3 x_4}\}} = 2, \\ \mathcal{I}nd[Y_1(x)]_{\{\overrightarrow{x_1 x_2} \cup \overrightarrow{x_3 x_4}\}} = -2, \end{cases} \tag{5.3.9}$$

The functions $Y_i(x)$, $i = 1, 2$, are meromorphic in the cut plane and their poles and zeros are the zeros of $a(x)$ and $c(x)$. Thus it follows from (5.3.9) that $Y_0(x)$ has two zeros and no pole and, conversely, that $Y_1(x)$ has two poles and no zero. On the other hand, from the proof of part (i), lemma 5.3.1 and the fact that

$$\left| \frac{Y_0(x)}{Y_1(x)} \right| \leq 1, \quad x \in \Gamma, \text{ used in particular at } x = 1 \text{ and } x = -1,$$

we have

$$\lim_{x \to \infty} \left| \frac{Y_0(x)}{Y_1(x)} \right| \leq 1, \quad \text{with equality only if } x_4 = \infty.$$

Applying now the maximum modulus principle to the function $\frac{Y_0(x)}{Y_1(x)}$, which is holomorphic in the complex plane cut along $[x_1 x_2] \cup [x_3 x_4]$, we get immediately the last assertion of (ii). The proof of theorem 5.3.3 is concluded.   ■

Corollary 5.3.5   1. $G_{\mathcal{M}} - [x_1 x_2] \xrightarrow{\frac{Y_0(x)}{X_0(y)}} G_{\mathcal{L}} - [y_1 y_2]$ *and the mappings are conformal.*

2. *The values of Y_0 belong to $G_{\mathcal{L}} \bigcup G_{\mathcal{L}_{ext}}$.*

3. *The values of Y_1 belong to $G_{\mathcal{L}}^c \bigcup G_{\mathcal{L}_{ext}}^c$.*

4. *The singular curves of the composed functions* $X_i \circ Y_j$ *are listed thereafter:*

$$X_0 \circ Y_0 \; : \; \mathcal{M} \bigcup [x_3 \bar{x}_4],$$

$$X_1 \circ Y_0 \; : \; \mathcal{M} \bigcup [x_1 x_2],$$

$$X_0 \circ Y_1 \; : \; \mathcal{M}_{ext} \bigcup [x_3 \bar{x}_4],$$

$$X_1 \circ Y_1 \; : \; \mathcal{M}_{ext} \bigcup [x_1, x_2].$$

Moreover, when $G_{\mathcal{M}} \subset G_{\mathcal{M}_{ext}}$, *the following automorphy relationships hold.*

$$X_0 \circ Y_0(t) = \begin{cases} t, & \text{if } t \in G_{\mathcal{M}}, \\ \neq t, & \text{if } t \in G_{\mathcal{M}}^c. \end{cases} \quad \text{Then} \quad X_0 \circ Y_0(G_{\mathcal{M}}^c) = G_{\mathcal{M}}.$$

$$X_1 \circ Y_0(t) = \begin{cases} t, & \text{if } t \in G_{\mathcal{M}}^c, \\ \neq t, & \text{if } t \in G_{\mathcal{M}}. \end{cases} \quad \text{Then} \quad X_1 \circ Y_0(G_{\mathcal{M}}) = G_{\mathcal{M}}^c.$$

$$X_0 \circ Y_1(t) = \begin{cases} t, & \text{if } t \in G_{\mathcal{M}_{ext}}, \\ \neq t, & \text{if } t \in G_{\mathcal{M}_{ext}}^c. \end{cases} \quad \text{Then} \quad X_0 \circ Y_1(G_{\mathcal{M}_{ext}}^c) = G_{\mathcal{M}_{ext}}.$$

$$X_1 \circ Y_1(t) = \begin{cases} t, & \text{if } t \in G_{\mathcal{M}_{ext}}^c, \\ \neq t & \text{if } t \in G_{\mathcal{M}_{ext}}. \end{cases} \quad \text{Then} \quad X_1 \circ Y_1(G_{\mathcal{M}_{ext}}) = G_{\mathcal{M}_{ext}}^c.$$

Proof. Assertion 1 follows directly from theorem 5.2.3 and so do 2 and 3, by application of the maximum modulus principle to the functions $Y_0(x)$ and $\dfrac{1}{Y_1(x)}$ respectively. The last assertion can be checked up to some tedious verifications which will be omitted. The proof of the corollary is terminated. ∎

Clearly, corollary 5.3.5 enables us to make the analytic continuation of the functions π and $\widetilde{\pi}$. This yields another derivation of theorem 3.2.2, starting from equation (5.1.2).

5.4 Index and Solution of the BVP (5.1.5)

As it emerges from section 5.2, a fundamental quantity in the solution of BVP's is the *index*. Problem (5.1.5) has been obtained from (5.1.3) and we have to combine the two basic constraints imposed on π and $\widetilde{\pi}$, i.e. they must be holomorphic inside their respective unit disk $\mathcal{D}$ and continuous on the boundary Γ (the unit circle). This is to say that in (5.1.5) π might be meromorphic in $G_{\mathcal{M}} \backslash \overline{\mathcal{D}}$ (see section 5.3 for the notation G_U). This leads to the notion of *reduced index*, which is the index of (5.1.5) subject to the constraints on π and $\widetilde{\pi}$ simultaneously. We shall proceed in three steps.

Notation All parameters have to be taken from (5.1.2). For any continuous non vanishing function $f(t)$ given on a contour $\mathcal{C}$, let $N_Z[f, G]$ denote the number of zeros of $f(t)$ in the open domain G having $\mathcal{C}$ as its boundary.

Theorem 5.4.1 *Let us introduce the two following quantities:*

$$
\delta \overset{def}{=}
\begin{cases}
0, \text{ if } Y_0(1) < 1 \text{ or } \left\{ Y_0(1) = 1 \text{ and } \dfrac{dq(x, Y_0(x))}{dx}\Big|_{x=1} > 0 \right\}, \\[12pt]
1, \text{ if } Y_0(1) = 1 \text{ and } \dfrac{dq(x, Y_0(x))}{dx}\Big|_{x=1} < 0.
\end{cases}
\tag{5.4.1}
$$

$$
\widetilde{\delta} \overset{def}{=}
\begin{cases}
0, \text{ if } X_0(1) < 1 \text{ or } \left\{ X_0(1) = 1 \text{ and } \dfrac{d\widetilde{q}(X_0(y), y)}{dy}\Big|_{y=1} > 0 \right\}, \\[12pt]
1, \text{ if } X_0(1) = 1 \text{ and } \dfrac{d\widetilde{q}(X_0(y), y)}{dy}\Big|_{y=1} < 0.
\end{cases}
\tag{5.4.2}
$$

Then the functional equation (5.1.4), with boundary condition (5.1.5), reduces to the BVP (5.4.19), the index of which is given by

$$
\widetilde{\chi} = -\mathcal{I}nd[K(t)]_{\mathcal{M}} = -L - M + \delta + \widetilde{\delta} - \mathbb{1}_{\{X_0(1)=1, Y_0(1)=1\}}.
\tag{5.4.3}
$$

Moreover, (5.1.1) admits a probabilistic solution if, and only if,

$$
\delta + \widetilde{\delta} = \mathbb{1}_{\{X_0(1)=1, Y_0(1)=1\}} + 1.
\tag{5.4.4}
$$

∎

Lemma 5.4.2

$$
\begin{aligned}
\widetilde{\chi_1} \overset{def}{=} & -\mathcal{I}nd[A(t)]_{\mathcal{M}} = -\mathcal{I}nd[q(x, Y_0(x))]_{\mathcal{M}} + \mathcal{I}nd[\widetilde{q}(X_0(y), y)]_{[\widehat{y_1 y_2}]} \\
= & \ \delta + \widetilde{\delta} - L - M - N_Z[q(x, Y_0(x)), G_{\mathcal{M}}] + N_Z[q(x, Y_0(x)), \mathcal{D}] \\
& + N_Z[\widetilde{q}(X_0(y), y), \mathcal{D}],
\end{aligned}
\tag{5.4.5}
$$

∎

Proof.   As announced in the beginning of section 5.3, the given functions $q(x, y)$ and $\widetilde{q}(x, y)$ are supposed to have suitable analytic continuations with respect to x [resp. y] in the domains $G_{\mathcal{M}}$ [resp. $G_{\mathcal{L}}$]. Then the first equality in (5.4.5) is a direct consequence of the definition of $A(t)$ and of the first *automorphy* property given in corollary 5.3.5. To establish (5.4.5) we shall compute separately the two terms coming in the right-hand side member, mainly using the principle of the argument. We have

$$
\mathcal{I}nd[\widetilde{q}(X_0(y), y)]_{[\widehat{y_1 y_2}]} = -M + N_Z[\widetilde{q}(X_0(y), y), \mathcal{D}] + \widetilde{\delta},
\tag{5.4.6}
$$

where $\widetilde{\delta}$ has been defined in (5.4.2). Equation (5.4.6) follows from the principle of the argument and from the properties of $X_0(y)$.

- When $\widetilde{\delta} = 0$, the same argument shows that, for $\varepsilon > 0$ sufficiently small,

$$N_Z[\widetilde{q}(X_0(y), y), \mathcal{D}] = N_Z[\widetilde{q}(X_0(y), y), \mathcal{D}_{1-\varepsilon}] = M,$$

because on $\Gamma_{1-\varepsilon}$, which denotes here the unit circle with a small indentation to the left of the point $y = 1$, we have

$$\Re\left[\frac{\widetilde{q}(X_0(y), y)}{y^M}\right] < 0, \text{ so that } \mathcal{I}nd\left[\frac{\widetilde{q}(X_0(y), y)}{y^M}\right]_{\Gamma_{1-\varepsilon}} = 0.$$

- When $\widetilde{\delta} = 1$, the same argument shows that

$$N_Z[\widetilde{q}(X_0(y), y), \mathcal{D}] = N_Z[\widetilde{q}(X_0(y), y), \mathcal{D}_{1+\varepsilon}] - 1 = M - 1,$$

using now the circle $\Gamma_{1+\varepsilon}$, having an indentation to the right of the point $y = 1$.

Thus (5.4.6) is proved.

Continuing with the proof of (5.4.5), we have

$$\mathcal{I}nd[q(x, Y_0(x))]_{\mathcal{M}} + \mathcal{I}nd[q(x, Y_0(x))]_{[\cancel{x_1 x_2}]} = N_Z[q(x, Y_0(x)), G_{\mathcal{M}}]. \tag{5.4.7}$$

Up to an exchange of the variables in (5.4.6), we can also write

$$\mathcal{I}nd[q(x, Y_0(x))]_{[\cancel{x_1 x_2}]} = -L + N_Z[q(x, Y_0(x)), \mathcal{D}] + \delta, \tag{5.4.8}$$

where δ has been defined in (5.4.1). Putting together (5.4.6), (5.4.7) and (5.4.8) we obtain (5.4.5) and the proof of lemma 5.4.2 is terminated. ∎

The function π is sought to be holomorphic in $\mathcal{D}$ and continuous in $\overline{\mathcal{D}}$, but might have poles in $G_{\mathcal{M}} \cap (\overline{\mathcal{D}})^c$. From corollary 5.3.5, we know that

$$|Y_0(x)| \leq 1, \text{ for } x \in G_{\mathcal{M}} \cap (\overline{\mathcal{D}})^c,$$

this being a consequence of the maximum modulus principle. Hence, rewriting (5.1.3) as

$$q(x, Y_0(x))\pi(x) + \widetilde{q}(x, Y_0(x))\widetilde{\pi}(Y_0(x)) + \pi_0(x, Y_0(x)) = 0,$$

one sees that the possible poles of $\pi(x)$ in $G_{\mathcal{M}} \cap (\overline{\mathcal{D}})^c$ are necessarily zeros of $q(x, Y_0(x))$ in this region, as $\widetilde{\pi}(y)$ is holomorphic in $\mathcal{D}$ and continuous in $\overline{\mathcal{D}}$. Setting

$$\begin{cases} S(x) &= \displaystyle\prod_{r_j \in G_{\mathcal{M}} \cap (\overline{\mathcal{D}})^c} (x - r_j), \\ \pi(x) &= \dfrac{\Phi(x)}{S(x)}, \quad G(x) = \dfrac{A(x)}{S(x)}, \end{cases} \tag{5.4.9}$$

where the r_j's stand for all the zeros of $q(x, Y_0(x))$ in $G_\mathcal{M} \cap (\overline{\mathcal{D}})^c$, the BVP (5.1.5) takes the form

$$\Phi(t)G(t) - \Phi(\alpha(t))G(\alpha(t)) = g(t), \quad t \in \mathcal{M} \qquad (5.4.10)$$

where now $\Phi(x)$ is required to be holomorphic in $G_\mathcal{M}$. Since

$$\mathcal{I}nd[G(x)]_\mathcal{M} = \mathcal{I}nd[A(x)]_\mathcal{M} - N_Z[q(x, Y_0(x)), G_\mathcal{M} \cap (\overline{\mathcal{D}})^c],$$

the index of (5.4.10) writes, according to (5.2.43) and (5.4.5),

$$\widetilde{\chi}_2 \overset{\text{def}}{=} -\mathcal{I}nd[G(x)]_\mathcal{M} = +\delta + \widetilde{\delta} - L - M - \mathbb{1}_{\{1 \in G_\mathcal{M}, Y_0(1)=1\}}$$

$$+ N_Z[q(x, Y_0(x)), \mathcal{D} \cap (G_\mathcal{M}^c)] + N_Z[\widetilde{q}(X_0(y), y), \mathcal{D}]. \qquad (5.4.11)$$

To proceed further with the *reduction* of the index, we must introduce two additional informations:

(i) First, whenever

$$\widetilde{q}(X_0(y), y) = 0, \quad y \in \overline{\mathcal{D}},$$

the product $\widetilde{q}(X_0(y), y)\widetilde{\pi}(y)$ taking place in (5.1.3) does vanish. Therefore, introducing

$$\psi_1(x) = \sum_k \frac{\pi(x) - \pi(X_0(u_k))}{x - X_0(u_k)}, \qquad (5.4.12)$$

where the summation is taken over the k's such that

$$\widetilde{q}(X_0(u_k), u_k) = 0, \quad u_k \in \mathcal{D},$$

(with the correct multiplicities), $\psi_1(x)$ is clearly holomorphic in the neighbourhood of $x = X_0(u_k)$. Moreover in (5.4.12) we have from (5.1.3),

$$\pi(X_0(u_k)) = \frac{-\pi_0(X_0(u_k), u_k)}{q(X_0(u_k), u_k)}, \qquad (5.4.13)$$

which gives the $\pi(X_0(u_k))$ in terms of the $\pi_0(X_0(u_k), u_k)$, i.e. linearly with respect to the $L + M - 1$ constants defining $\pi_0(x, y)$ [see equation (5.1.2)].

(ii) Secondly, to take into account the zeros of $q(x, Y_0(x))$ in the region $\mathcal{D} \cap G_\mathcal{M}^c$, one can start from (5.1.4), which, by analytic continuation, is indeed valid for all y in the complex plane. In particuler, it is possible to replace in (5.1.4) y by $Y_0(x)$, so that (5.1.4) takes the form

$$\pi(X_0 \circ Y_0(x))A(X_0 \circ Y_0(x)) - \pi(x)A(x) = g(x), \quad \text{for } x \in \overline{\mathcal{D}} \cap G_\mathcal{M}^c. \qquad (5.4.14)$$

To obtain (5.4.14) we have replaced in (5.1.4) y by $Y_0(x)$ and have taken into account the automorphy relation of corollary 5.3.5 $X_1 \circ Y_0(x) = x$, for $x \in G_\mathcal{M}^c$. Thus, since

$$A(x) = \frac{q(x, Y_0(x))}{\bar{q}(x, Y_0(x))},$$

we get from (5.4.14) the additional relationships

$$\pi(X_0 \circ Y_0(v_k)) = \frac{g(v_k)}{A(X_0 \circ Y_0(v_k))}, \tag{5.4.15}$$

for any $v_k \in \mathcal{D} \cap G_{\mathcal{M}}^c$, such that $q(v_k, Y_0(v_k)) = 0$. Then, proceeding as in (5.4.12), we set

$$\begin{aligned}
\psi(x) &= \psi_1(x) + \sum_{\ell} \frac{\pi(x) - \pi(X_0 \circ Y_0(v_\ell))}{x - X_0 \circ Y_0(v_\ell)} \\
&= \sum_k \frac{\pi(x) - \pi(X_0(u_k))}{x - X_0(u_k)} + \sum_k \frac{\pi(x) - \pi(X_0 \circ Y_0(v_\ell))}{x - X_0 \circ Y_0(v_\ell)}.
\end{aligned} \tag{5.4.16}$$

Inverting now (5.4.16) yields

$$\pi(x) = \psi(x)\frac{R(x)}{R'(x)} + T(x), \tag{5.4.17}$$

where, using (5.4.13) and (5.4.15),

$$\begin{cases}
R(x) = \prod_{k,\ell}(x - X_0(u_k))(x - X_0 \circ Y_0(v_\ell)), \\
T(x) = \dfrac{R(x)}{R'(x)}\left[\displaystyle\sum_\ell \frac{g(v_\ell)}{(x - X_0 \circ Y_0(v_\ell))A(X_0 \circ Y_0(v_\ell))} - \sum_k \frac{\pi_0(X_0(u_k), u_k)}{q(X_0(u_k), u_k)}\right].
\end{cases} \tag{5.4.18}$$

Joining together (5.4.9), (5.4.10), (5.4.17) and (5.4.18), one does obtain the final *reduced* BVP

$$\boxed{\rho(t)K(t) - \rho(\alpha(t))K(\alpha(t)) = k(t), \quad t \in \mathcal{M},} \tag{5.4.19}$$

with

$$\begin{cases}
\rho(t) = \dfrac{\Phi(t)R(t)}{R'(t)}, \quad K(t) = \dfrac{G(t)R'(t)}{R(t)}, \\
k(t) = g(t) + T(\alpha(t))A(\alpha(t)) - T(t)A(t), \\
\alpha(t) = \bar{t}.
\end{cases} \tag{5.4.20}$$

In (5.4.19), ρ is sought to be holomorphic in $G_{\mathcal{M}}$, and all the assumptions concerning π and $\tilde{\pi}$ have been used. This explains the expression *final reduced index* employed above.

Existence and uniqueness of a solution of (5.4.19) are strictly equivalent to the ergodicity of the random walk defined in chapter 1, the invariant measure of which satisfies (5.1.1). According to (5.2.43), the index of (5.4.19) is given by

$$\mathcal{I}nd[K(t)]_{\mathcal{M}} = \mathcal{I}nd[G(t)]_{\mathcal{M}} + N_Z[R(x), G_{\mathcal{M}}]$$
$$= \mathcal{I}nd[G(t)]_{\mathcal{M}} + N_Z[q(x, Y_0(x)), \mathcal{D} \cap G_{\mathcal{M}}^c] + N_Z[\tilde{q}(X_0(y), y), \mathcal{D}],$$

which, using (5.4.11), is exactly the value announced in (5.4.3). The first part of theorem 5.4.1 is proved.

To derive (5.4.4), note first that, since $L, M \geq 1$, (5.4.3) ensures $\tilde{\chi} \leq -1$. Thus, by theorem 5.2.8, a solution of (5.4.19) exists and is unique if, and only if,

$$L + M - \delta - \tilde{\delta} + \mathbb{1}_{\{X_0(1)=1, Y_0(1)=1\}} - 1$$

conditions of the form (5.2.13) are satisfied. These conditions are linear and involve only the $L + M - 1$ unknowns coming in the definition of $\pi_0(x, y)$. Remarking that the existence (or non-existence) of an invariant measure for a Markov chain is not subject to the modification of a *finite* number of parameters (e.g. transition probabilities) of the chain, one can always assume that the conditions are independent. Taking also into account the normalizing condition, (which says that the invariant distribution is *proper*), we will get a non-homogeneous linear system of $L + M - 1$ equations with $L + M - 1$ unknowns, if, and only if, equation (5.4.4) holds. Moreover, lemma 2.2.1 applies. The proof of theorem 5.4.1 is concluded. ∎

Theorem 5.4.3 *Under the condition (5.4.4), the function π is given by*

$$\pi(x) = \frac{R'(x)H(x)}{2i\pi R(x)S(x)} \int_{\mathcal{M}_d} \frac{k(t)w'(t)dt}{H^+(t)K(\bar{t})(w(t) - w(x))}, \quad \forall x \in G_{\mathcal{M}}. \quad (5.4.21)$$

where

 (i) *$\mathcal{M}_d$ denotes the portion of the curve $\mathcal{M}$ located in the lower half-plane $\Im z \leq 0$;*

 (ii) *k and K have been introduced in (5.4.20);*

 (iii) *w is solution of the BVP (5.2.39) on the curve $\mathcal{M}$ (see theorem 5.2.7);*

 (iv)

$$H(t) = (w(t) - X_0(y_2))^{-\tilde{\chi}} e^{\Gamma(t)}, \quad t \in G_{\mathcal{M}},$$

$$\Gamma(t) = \frac{1}{2i\pi} \int_{\mathcal{M}_d} \log \frac{K(\bar{s})}{K(s)} \frac{w'(s)ds}{w(s) - w(t)}, \quad t \in G_{\mathcal{M}},$$

$$H^+(t) = (w(t) - X_0(y_2))^{-\tilde{\chi}} e^{\Gamma^+(t)}, \quad t \in \mathcal{M}_d,$$

$$\Gamma^+(t) = \frac{1}{2} \log \frac{K(\bar{t})}{K(t)} + \frac{1}{2i\pi} \int_{\mathcal{M}_d} \log \frac{K(\bar{s})}{K(s)} \frac{w'(s)ds}{w(s) - w(t)}, \quad t \in \mathcal{M}_d.$$

Proof.   Just a direct consequence of (5.2.44). ∎

5.5 Complements

5.5.1 Analytic Continuation

In fact the functional equation (5.1.5) can in turn be used to make the analytic continuation of π, in a finite number of steps, to the whole complex plane cut along $[x_3 x_4]$. Without going into details, remark just that the process consists in making use of the automorphy relationships established in corollary 5.3.5. Consequently, it is also possible to set a BVP on the curve $\mathcal{M}_{ext}$, provided that the generating functions of the jumps on the axes have suitable continuations. An example of this possibility is given in [27].

5.5.2 Computation of w

There are three main possible ways to obtain the function w, which realizes the conformal mapping of $G_{\mathcal{M}}$ onto the complex plane cut along an arc.

5.5.2.1 An Explicit Form via the Weierstrass $\wp$-Function.

Any couple (x, y) solution of the algebraic equation $Q(x, y) = 0$ corresponds to exactly one point s on the Riemann surface $\mathbf{S}$. This mapping will be denoted by $(x(s), y(s))$. We refer indeed to the uniformization of the algebraic curve, proposed in chapter 3. It turns out that the surface $\mathbf{S}$ can be viewed as a semi-open rectangle $[0, \omega_2[\times [0, \omega_1[$, representing the algebraic function $Y(x)$ and shown in figure 5.5.1.

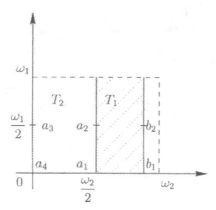

Fig. 5.5.1.

The affixes of $a_i, b_i, i = 1, \ldots, 4$ have been determined in chapter 3: they correspond to (3.1.7), (3.1.8), (3.1.9), and we know that

$$\omega_3 = 2(b_1 - a_1).$$

The sheet $\mathbf{T}_1$ in figure 5.5.1 is that of the branch $Y_0(x)$, since it contains the point b_1, which on $\mathbf{S}$ corresponds to $(X(y_1), Y_0(X(y_1)))$, since

$$Y_0(X(y_i)) = y_i, \quad i = 1, 2.$$

We are looking for the function w (coming in the solution of the BVP presented in the preceding sections), which is defined in the complex plane $\mathbb{C}_x$, holomorphic inside $G_{\mathcal{M}}$, except at one point where it has a simple pole, and subject to the *gluing* condition

$$w(x) = w(\overline{x}), \quad \forall x \in \mathcal{M}. \tag{5.5.1}$$

For $x \equiv x(\omega)$, the homographic function of $\wp$ in (3.3.3), we shall write $w \circ x \overset{\text{def}}{=} \widetilde{w}$. The problem for $\widetilde{w}$ can be stated as follows. *Find a function $\widetilde{w}$ holomorphic in the shaded area represented in figure 5.5.1, in the interior of $\mathbf{T}_1$, having one pole (which can be chosen arbitrary, but not on the segment $[b_1, b_1 + \omega_1[$), and satisfying the next two conditions:*

(i) For $\omega \in [b_1, b_1 + \omega_1[$, $\widetilde{w}$ *glues the two edges of the cut* $[y_1, y_2]$, *which reads*

$$\widetilde{w}\left(b_1 + \frac{\omega_1}{2} + u\right) = \widetilde{w}\left(b_1 + \frac{\omega_1}{2} - u\right), \quad u \in \left]-\frac{\omega_1}{2}, \frac{\omega_1}{2}\right[\tag{5.5.2}$$

and is nothing else but (5.5.1);

(ii) $\widetilde{w}$ *also glues the edges of the cut* $[x_1, x_2]$, *which is located inside* $G_{\mathcal{M}}$ *in* $\mathbb{C}_x$. *Thus*

$$w^+(x) = w^-(x), \quad \forall x \in [x_1, x_2],$$

or, equivalently,

$$\widetilde{w}(a_2 + u) = \widetilde{w}(a_2 - u), \quad u \in \left]-\frac{\omega_1}{2}, \frac{\omega_1}{2}\right[. \tag{5.5.3}$$

Let $\psi(\omega) \overset{\text{def}}{=} \widetilde{w}(a_2 + \omega)$. Since $b_1 + \frac{\omega_1}{2} = a_2 + \frac{\omega_3}{2}$, the two conditions (5.5.2) and (5.5.3) can be rewritten as

$$\psi(\omega) = \psi(-\omega), \tag{5.5.4}$$
$$\psi\left(\omega + \frac{\omega_3}{2}\right) = \psi\left(-\omega + \frac{\omega_3}{2}\right), \tag{5.5.5}$$

where ψ has to be holomorphic in $\left]\frac{-\omega_1}{2}, \frac{\omega_1}{2}\right[\times [0, \frac{\omega_3}{2}[$. Equation (5.5.4) permits to continue ψ to the rectangle $\left]\frac{-\omega_1}{2}, \frac{\omega_1}{2}\right[\times [\frac{-\omega_3}{2}, 0]$, as a meromorphic function, so that ψ is even in $\left]\frac{-\omega_1}{2}, \frac{\omega_1}{2}\right[\times [\frac{-\omega_3}{2}, \frac{\omega_3}{2}[$, where it has two simple poles symmetric with respect to the origin. Hence, using (5.5.5), we can write

$$\psi(u) = \psi(u + \omega_1) = \psi(u + \omega_3), \quad \forall u \in \,]\frac{-\omega_1}{2}, \frac{\omega_1}{2}[\times [\frac{-\omega_3}{2}, \frac{\omega_3}{2}[. \qquad (5.5.6)$$

Now the relations given in (5.5.6) allow to continue ψ to the whole complex plane $\mathbb{C}$, as a doubly-periodic meromorphic function $\psi(u)$, with periods ω_1 and ω_3. Moreover, in each rectangle, the two poles can be taken to coincide at the center $u = 0$ of such rectangle. Finally, we have proved that an admissible choice for ψ is simply

$$\psi(u) = \wp(u; \omega_1, \omega_3).$$

5.5.2.2 A Differential Equation. The results obtained in the preceding subsection yield directly

$$\frac{d\widetilde{w}}{d\omega} = \wp'\left(\omega - \frac{\omega_1 + \omega_2}{2}; \omega_1, \omega_3\right)$$

$$= \sqrt{(\widetilde{w}(\omega) - e_1)(\widetilde{w}(\omega) - e_2)(\widetilde{w}(\omega) - e_3)},$$

where

$$e_1 = \wp\left(\frac{\omega_2}{2}; \omega_1, \omega_3\right), \quad e_2 = \wp\left(\frac{\omega_2 + \omega_3}{2}; \omega_1, \omega_3\right), \quad e_3 = \wp\left(\frac{\omega_1 + \omega_2 + \omega_3}{2}; \omega_1, \omega_3\right).$$

Also,

$$\frac{dx}{d\omega} = \frac{C}{\sqrt{D(x)}},$$

where $D(x)$ is the discriminant in the uniformization formulas (see chapter 3), so that, choosing $C = 1$, w satisfies the differential equation

$$\frac{dw}{dx} = \frac{\sqrt{(w - e_1)(w - e_2)(-e_3)}}{\sqrt{D(x)}}.$$

5.5.2.3 An Integral Equation. According to theorem 5.2.7, we know that w is given by a Cauchy type integral (5.2.38), the density of which satisfies the quasi-Fredholm integral equation (5.2.35). The derivative $\alpha'(t)$, which comes in the kernel of (5.2.35), can be expressed explicitly, noting that

$$\alpha(t) = X_1 \circ Y_0(t), \quad t \in \mathcal{M}.$$

Skipping over the details, we obtain

$$\alpha'(t) \equiv \frac{d\alpha(t)}{dt} = e^{i[\pi - \arg(b^2(t) - 4a(t)c(t))]}, \quad t \in \mathcal{M},$$

where the functions a, b and c are the second degree polynomials of section 5.3.

6. The Genus 0 Case

This chapter is devoted to the case when the algebraic curve defined by the equation $Q(x, y) = 0$ has genus 0. The corresponding classification was made in Chapter 2 and exactly five situations have been found, described by the relations (2.3.5) to (2.3.8). In fact, since (2.3.6) and (2.3.8) are equivalent up to a permutation of the variables x and y, we are left with four significantly different cases, which will be treated separately.

In some sense, genus 0 can be viewed, in the complex plane, as a degenerate limiting case of genus 1. Thus many results established in section 5.3 about the branches of the algebraic functions X and Y still hold, in particular most of the properties state i, (ii) and (iii) of theorem 5.3.3. It is worth recalling (see remark 2.5.2) that the analytic continuation process could also be carried out on the Riemann sphere, as was done in chapter 2 on the torus in the genus 1 case. Nevertheless, as a matter of continuity with respect to chapter 5, we choose to make the complete analysis in the complex plane, since it presents some new interesting features pertaining in particular to automorphic functions [33].

6.1 Properties of the Branches

The main properties are summarized in the next theorem.

Theorem 6.1.1
(i) The algebraic curve defined by $Q(x, y) = 0$ has genus 0 in the following cases.

1. $\begin{cases} x_1 = x_2 = 0 \\ y_3 = y_4 = \infty. \end{cases}$

2. $\begin{cases} y_1 = y_2 = 0 \\ x_3 = x_4 = \infty. \end{cases}$

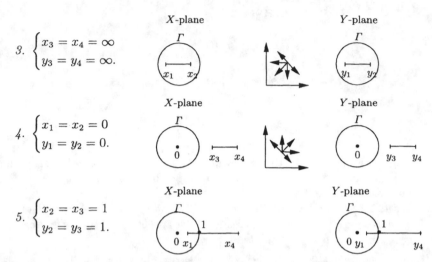

3. $\begin{cases} x_3 = x_4 = \infty \\ y_3 = y_4 = \infty. \end{cases}$

4. $\begin{cases} x_1 = x_2 = 0 \\ y_1 = y_2 = 0. \end{cases}$

5. $\begin{cases} x_2 = x_3 = 1 \\ y_2 = y_3 = 1. \end{cases}$

In addition x_3 and y_3 are always positive, but x_4 and y_4 need not be positive. If for instance $x_4 < 0$, then the plane is cut along $[-\infty, x_4] \cup [x_3, +\infty]$.

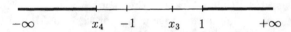

(ii) The cases listed in (i) above correspond respectively to the relations (2.3.9), (2.3.7), (2.3.6), (2.3.8) and (2.3.5) and the following more precise properties hold.

- The branches X_i and Y_i, $i = 1, 2$, are meromorphic in their respective complex plane cut along a single slit as shown in part (i) of the theorem.

- $|X_0(y)| \le |X_1(y)|$ [resp. $|Y_0(x)| \le |Y_1(x)|$] in the whole complex plane properly cut. Equalities can take place only at $0, \infty$ or on the cuts.

- In case 1, Y_0, Y_1 [resp. X_0, X_1] have a common pole [resp. zero] at $x = 0$ [resp. $y = \infty$] and X_0 is bounded.

- In case 2, Y_0, Y_1 [resp. X_0, X_1] have a common zero [resp. pole] at $x = \infty$ [resp. $y = 0$] and Y_0 is bounded.

- In case 3, Y_0, Y_1 [resp. X_0, X_1] have a common pole at $x = \infty$ [resp. $y = \infty$].

- In case 4, Y_0, Y_1 [resp. X_0, X_1] have a common zero at $x = 0$ [resp. $y = 0$]. The functions X_0 and Y_0 are bounded, but X_1 and Y_1 have two poles.

- In case 5, Y_0 [resp. X_0] is holomorphic in the plane cut along $[x_1 x_4]$ (resp. $[y_1 y_4]$). The branches Y_1 and X_1 have two poles.

■

Proof. All statements of part *(i)* follow directly from section 2.3. The sharper results of *(ii)* can be derived exactly (even in a simpler way) along the lines of lemma 5.3.1 and theorem 5.4.1, using the separation of the branches on the unit circle, together with the principle of the argument and the maximum modulus theorem. The details are omitted. ∎

Now we shall analyze in detail the various possibilities listed above. To avoid uninteresting technicalities, the functions π_0, q and $\widetilde{q}$ in (5.1.2) will be supposed to be polynomials with respect to the two variables (x, y). This amounts to say that the jumps of the random walk on the axes are bounded. On the other hand, the ergodicity conditions are again given by condition (5.4.4), by direct continuity with respect to the parameters $\{p_{i,j}\}$, except perhaps for the case 5.

6.2 Case 1: $p_{01} = p_{-1,0} = p_{-1,1} = 0$

Theorem 6.2.1 *Under the above conditions, the functions π and $\widetilde{\pi}$ exist if, and only if, (5.4.4) holds. In this case the following situation holds:*

- *$\widetilde{\pi}$ is a rational function and its poles in the complex plane are the zeros of $\widetilde{q}(X_0(y), y)$ located in G_Γ^c (i.e. outside the unit disc).*

- *π has the form*
$$\pi(x) = U(x)Y_0(x) + V(x),$$
where U and V are rational functions. ∎

Proof. As in chapter 5, it is possible to state a BVP of the form (5.1.5), for the function π, on a curve which was denoted $\mathcal{M}$ in section 5.3. The computation of the index of this BVP can be derived exactly as in theorem 5.4.1 and the ergodicity conditions are still given by (5.4.4). When they hold, it is however not necessary to solve this BVP, as there is a much more direct way to obtain the solution. In fact, writing
$$\widetilde{\pi}(y) = \frac{-q(X_0(y), y)\pi(X_0(y)) - \pi_0(X_0(y), y)}{\widetilde{q}(X_0(y), y)}, \ \forall y \in G_\Gamma^c,$$
we get immediately the analytic continuation of $\widetilde{\pi}$ to the region G_Γ^c, since from the maximum modulus principle and by theorem 6.1.1, we have
$$|X_0(y)| \leq 1, \ \forall y \in G_\Gamma^c.$$
Moreover, $\forall y \in G_\Gamma^c$ the functions $q(X_0(y), y)$, $\widetilde{q}(X_0(y), y)$ and $\pi_0(X(y), y)$ are analytic and of finite degree at infinity, so that the only singularities of $\widetilde{\pi}$ in the whole complex plane are the potential zeros of $\widetilde{q}(X_0(y), y)$ in G_Γ^c and $\widetilde{\pi}$ is rational.

When $y \to \infty$, $\widetilde{\pi}(y) = \mathcal{O}(y^t)$ where $t = \max(r, s)$, with
$$\frac{q(X_0(y), y)}{\widetilde{q}(X_0(y), y)} = \mathcal{O}(y^r), \qquad \frac{\pi_0(X_0(y), y)}{\widetilde{q}(X_0(y), y)} = \mathcal{O}(y^s).$$
The proof of the theorem is concluded. ∎

6.3 Case 3: $p_{11} = p_{10} = p_{01} = 0$

Theorem 6.3.1 *Under the above conditions, the functions π and $\widetilde{\pi}$ exist if, and only if, (5.4.4) holds, in which case they can be continued as meromorphic functions, to the whole complex plane. In the domain $G_{\mathcal{M}}$, π has the integral representation given by (5.4.21), where $\mathcal{M}$ is an ellipse, which in the x-plane is given by the equation*

$$\frac{(u - u_0)^2}{a^2} + \frac{v^2}{b^2} = 1, \quad \text{with} \quad x = u + iv.$$

The function w, appearing in (5.4.21) and solution of (5.2.39), can be chosen as

$$w(z) = \frac{1}{2}\left(h(z) + \frac{1}{h(z)}\right), \quad \forall z \in G_{\mathcal{M}}, \tag{6.3.1}$$

where

$$h(z) = \sqrt{k(\rho)} \, \text{sn}\left(\frac{2K \arcsin(z - u_0)}{\pi}; \rho\right), \quad \text{with} \quad \rho = \left(\frac{a - b}{a + b}\right)^2,$$

represents the conformal mapping of the interior of the above ellipse onto the unit disc, $k(.)$ and K being respectively the modulus and the real quarter period of the Jacobi elliptic function $\text{sn}\,(z; k)$. Similar conclusions hold for $\widetilde{\pi}$. ∎

Proof.   To show the first and third claims of the theorem, it suffices to note that the general analysis of chapter 5 applies *verbatim*, the verification that $\mathcal{L}$ and $\mathcal{M}$ are ellipses being elementary. From this integral formula, it is in fact possible to get the analytic continuation of π to the whole complex plane. Nonetheless, it is instrumental to prove the second point of the theorem directly from the functional equations (5.1.3) or (5.1.5), using some properties of the branches Y_0 and Y_1 given in the next lemma, which is the analogue of corollary 5.3.5.

Lemma 6.3.2

$$|X_1(t)| \geq \frac{|\widetilde{b}(t)|}{2} \quad \text{and} \quad |Y_1(t)| \geq \frac{|b(t)|}{2}, \, \forall t.$$

$$X_0 \circ Y_1(t) = t \quad \text{and} \quad |X_1 \circ Y_1(t)| > |t|, \forall t.$$

$$X_0 \circ Y_0(t) = \begin{cases} t, & \text{if } t \in G_{\mathcal{M}}, \\ \neq t, & \text{if } t \in G_{\mathcal{M}}^c \end{cases} \quad \text{and} \quad X_0 \circ Y_0(G_{\mathcal{M}}^c) = G_{\mathcal{M}}.$$

$$X_1 \circ Y_0(t) = \begin{cases} t, & \text{if } t \in G_{\mathcal{M}}^c, \\ \neq t, & \text{if } t \in G_{\mathcal{M}} \end{cases} \quad \text{and} \quad X_1 \circ Y_0(G_{\mathcal{M}}) = G_{\mathcal{M}}^c.$$

Proof.   It relies on standard applications of the maximum modulus principle. The details are omitted. ∎

Lemma 6.3.3 *Let Δ_n the sequence of ring-shaped domains constructed recursively as follows:*

$$\Delta_{n+1} = X_1 \circ Y_1(\Delta_n), \forall n \geq 0, \qquad (6.3.2)$$

where Δ_0 is the doubly connected domain, with boundary the slit $[x_1 x_2]$ and the closed curve $X_1 \circ Y_1([x_1 x_2])$. Thus

$$\Delta_0 = G_{X_1 \circ Y_1([x_1 x_2])} - [x_1 x_2].$$

Let $\mathcal{D}_n \overset{def}{=} \bigcup_{k \leq n} \Delta_n$. Then

(i) $\mathcal{D}_{n+1} = \mathcal{D}_n \oplus \Delta_{n+1}$,   where $\oplus$ denotes the direct sum of sets;

(ii) $\lim_{n \to \infty} \mathcal{D}_n = \mathbb{C} - [x_1 x_2]$.

Proof. By induction. Let $[G]$ denote the boundary of an arbitrary domain G. Assume (i) holds up to some $n > 0$, but not for $n + 1$. Remark first that the *internal* boundary of Δ_{n+1} coincides with the *external* boundary of Δ_n. From the principle of correspondence of the boundaries for conformal mappings, there exist 3 points z_{n-1}, z_n, t_n, such that

$$z_n = X_1 \circ Y_1(t_n), \quad z_n, t_n \in [\mathcal{D}_n],$$
$$z_n = X_1 \circ Y_1(z_{n-1}), \quad z_{n-1}, \in [\mathcal{D}_{n-1}].$$

But in this case, it follows from lemma 6.3.2 that necessarily $t_n = z_{n-1}$, which contradicts $\mathcal{D}_{n-1} \subset \mathcal{D}_n$. So, to prove (i), it suffices to check the initial step, which is indeed straightforward. The point (ii) is also immediate by the first two properties of the branches listed in lemma 6.3.2. ■

Rewriting for the sake of completeness the basic twin equations

$$q(X_0(y), y)\pi(X_0(y)) + \tilde{q}(X_0(y), y)\tilde{\pi}(y) + \pi_0(X_0(y), y) = 0, \; \forall y \in \mathbb{C}, \quad (6.3.3)$$
$$q(x, Y_0(x))\pi(x) + \tilde{q}(x, Y_0(x))\tilde{\pi}(Y_0(x)) + \pi_0(x, Y_0(x)) = 0, \quad \forall x \in \mathbb{C}, \quad (6.3.4)$$

we can proceed by induction, in a flip-flop way as follows.

Assumption: π is meromorphic in $\mathcal{D}_n$.

Conclusion:

 1. $\tilde{\pi}$ is meromorphic in $Y_1(\mathcal{D}_n)$;
 2. π has a meromorphic continuation to $\mathcal{D}_{n+1}$.

Assertion 1 of the conclusion is obtained from equation (6.3.3), since the functions Y_1 and Y_2 are analytic outside the unit disk, except at $y = \infty$, where they have a single pole; assertion 2 follows by exploiting (6.3.4) and lemma 6.3.3.

The recursive computation of the poles rely on the forthcoming lemma, which is also another way of getting the meromorphic continuation of the various functions.

Lemma 6.3.4 *Equation (5.1.5) or its reduced form (5.4.19) can be expressed as*

$$\pi(u)f(u, Y_1(u)) - \pi(X_1 \circ Y_1(u))f(X_1 \circ Y_1(u), Y_1(u)) = g(u), \quad u \in [x_1 \, x_2], \tag{6.3.5}$$

which gives the meromorphic continuation of π to the whole complex plane.

Proof. From the analysis carried out in the preceding chapter, equation (5.1.4)

$$\pi(X_0(y))f(X_0(y), y) - \pi(X_1(y))f(X_1(y), y) = h(y), \ \forall y \in [y_1 y_2],$$

can be continued to $\mathbb{C}$. Writing it in particular for $y \in \mathcal{L}$, it is permitted, using lemma 6.3.2, to make the change of variables $y = Y_1(u)$, which implies exactly (6.3.5). ∎

Let

$$n_0 = \min_{n \geq 0}\{q(v, Y_0(v)) \, \widetilde{q}(v, Y_1(v)) \neq 0, \ \forall v \in \mathcal{D}_n^c\},$$

and $\alpha_0^k, \ k = 1, \dots, m$, be the poles of π in $\mathcal{D}_{n_0}$. Note that n_0 is finite, since we have assumed that q and $\widetilde{q}$ are polynomials. Then the possible poles of π can be recursively computed from the sequences $\{\alpha_n^k, 1 \leq k \leq m, \ n \geq 1\}$, where

$$\alpha_{n+1}^k = X_1 \circ Y_1(\alpha_n^k).$$

Similar results hold for $\widetilde{\pi}$, up to an exchange of parameters. ∎

6.4 Case 4: $p_{-1,0} = p_{0,-1} = p_{-1,-1} = 0$

Due to the position of the slit $[x_3, x_4]$, the problem here is of a slightly different nature and is connected with the automorphic functions (see [33]), as enlighted in the next theorem.

Theorem 6.4.1 *The functional equation*

$$\pi\big(X_0 \circ Y_0(t)\big)f\big(X_0 \circ Y_0(t), Y_0(t)\big) - \pi(t)f\big(t, Y_0(t)\big) = g(t) \tag{6.4.1}$$

is valid for all $t \in \mathbb{C}$ and provides the analytic continuation of π as a meromorphic function (the number of poles being finite) to the whole complex plane cut along $[x_3 \, x_4]$. ∎

Proof. By an informal topological argument, one can say that the cut $[y_1 y_2]$ has in some sense *shrunk* into a single point, the origin, where the algebraic function $Y(x)$ has a double zero. Hence, equation (5.1.4)

$$\pi(X_0(y))f(X_0(y), y) - \pi(X_1(y))f(X_1(y), y) = h(y),$$

obtained by elimination of $\widetilde{\pi}$, holds in a neighborhood of the origin in the Y-plane. The result of the theorem follows from the properties of the branches X_i, Y_i, listed in theorem 6.1.1. ∎

At once, it is important to note that (5.1.4), quoted above, cannot be continued to $[y_3, y_4]$, so that we do not have a BVP on a single curve, but on two curves. In the rest of this section, several methods are proposed to solve (6.4.1).

6.4.1 Integral Equation

We write π in the form

$$\pi(z) = \frac{1}{2\pi} \int_{[x_3 x_4]} \frac{\omega(t)dt}{t - z} + R(z), \quad \forall z \in \mathbb{C} - [x_3 x_4], \qquad (6.4.2)$$

where R is rational fraction having the same poles and residues as π. Letting now z go to the slit $[x_3 x_4]$ and using formula (5.2.4) in (6.4.1), we get

$$\omega(t) - \frac{1}{\pi} \int_{[x_3 x_4]} \frac{[u K_1(t) + K_2(t)]\omega(u)du}{|u - X_0 \circ Y_0(t)|^2} = K_3(t), \qquad (6.4.3)$$

where $K_i, i = 1, 2, 3$, are real functions, given in terms of f and g, their explicit expression being omitted. Conspicuously, the operator (6.4.3) is strongly contractant and can thus lead to an efficient numerical evaluation, all the more because the unknown density $\omega(t)$ is real.

6.4.2 Series Representation

It follows from the maximum modulus principle that the function $X_0 \circ Y_0$ satisfies

$$\begin{cases} |X_0 \circ Y_0(t)| \le |t|, & \forall t \in \mathbb{C}, \\ |X_0 \circ Y_0(t)| < |t|, & \forall t \in \mathcal{D}, \end{cases}$$

and admits 0 and 1 as fixed points. Thus, denoting by $X_0 \circ Y_0^{(n)}$ the n-th iterate of $X_0 \circ Y_0$, we have

$$\lim_{n \to \infty} X_0 \circ Y_0^{(n)}(t) = 0, \quad \forall t \in \mathcal{D}.$$

Consequently, by (6.4.1), $\pi(z)$ can be expressed as a series, for $z \in \mathcal{D}$, the terms of which are finite products of analytic functions.

6.4.3 Uniformization

The algebraic curve $Q(x, y) = 0$ is of genus 0 and, consequently, admits a rational uniformization, which will be expressed as

$$\begin{cases} x(s) &= \dfrac{x_3 + x_4}{2} + \dfrac{x_4 - x_3}{4}\left(s + \dfrac{1}{s}\right), \\[3mm] y(u) &= \dfrac{y_3 + y_4}{2} + \dfrac{y_4 - y_3}{4}\left(\eta(s) + \dfrac{1}{\eta(s)}\right) \end{cases} \qquad (6.4.4)$$

where the automorphisms ξ, η, δ, introduced in chapter 3, take the form

$$\begin{cases} \xi(s) = \dfrac{1}{s}, \\[2mm] \eta(s) = \dfrac{as+b}{cs-a}, \\[2mm] \delta(s) = \dfrac{bs+a}{-as+c}. \end{cases}$$

The function $X_0 \circ Y_0$ coming in equation (6.4.1) is, up to the above uniformization, strongly related to $\delta = \eta \circ \xi$. More information in given in the next lemma.

Lemma 6.4.2 *Let u_1, u_2 be the two fixed points of $\delta(s)$ in the s-plane. According to [33], we have the following classification:*

(1)   If $p_{10}^2 - 4p_{11}p_{1,-1} \neq 0$, then

$$\frac{\delta(s) - u_1}{\delta(s) - u_2} = K \frac{s - u_1}{s - u_2},$$

where u_1, u_2 are the two roots of the equation

$$(x_4 - x_3)u^2 + 2(x_4 + x_3)u + x_4 - x_3 = 0,$$

with $u_1 u_2 = 1$, and the multiplier K is given by

$$K = \frac{\eta(0) - u_1}{\eta(0) - u_2}.$$

(1.a)   If $p_{10}^2 - 4p_{11}p_{1,-1} > 0$, then x_4 is positive and

$$u_2 < -1 < u_1 < 0.$$

Moreover,

- *if $y_4 > 0$, then K is real, $|K| \neq 1$, so that $\delta(s)$ is of hyperbolic type;*
- *if $y_4 < 0$, then K is complex, $|K| \neq 1$, so that $\delta(s)$ is of loxodromic type.*

(1.b) If $p_{10}^2 - 4p_{11}p_{1,-1} < 0$, then x_4 is negative, u_1, u_2 are complex conjugate and located on the unit circle. In addition,

- *if $y_4 > 0$, then K is complex, $|K| = 1$, so that $\delta(s)$ is of elliptic type;*
- *if $y_4 < 0$, then K is complex, $|K| \neq 1$, so that $\delta(s)$ is again of loxodromic type.*

(2)   If $p_{10}^2 - 4p_{11}p_{1,-1} = 0$, then $x_4 = \infty$. The two fixed points coincide, with $u_1 = u_2 = -1$, and the uniformization of $x(s)$ in (6.4.2) becomes

$$x(s) = x_3 - s^2.$$

The transformation $\delta(s)$ is of parabolic type, and

$$\frac{1}{\delta(s)+1} = \frac{1}{s+1} + C.$$

∎

Proof.   The classification in the lemma relies on classical properties of the group generated by fractional linear transforms.   ∎

The above uniformization gives a representation of π as a series of meromorphic functions, the form of which depends on the type of δ, as defined in lemma 6.4.2.

6.4.4 Boundary Value Problem

Via (5.4.6), the slit $[x_3 x_4]$ in the X-plane is mapped onto the unit circle Γ in the z-plane. Similarly, the curve $X_0 \circ Y_0\left([\overleftrightarrow{x_3, x_4}]\right)$ corresponds to two circles in the z-plane, one of them, say $\mathcal{C}$, is located inside Γ. It is now possible to pose a problem of Carleman (see 5) for π in the annulus between $\mathcal{C}$ and Γ. Moreover, explicit expression do exist for the conformal mapping of this annulus onto simpler regions, allowing to transform the Carleman BVP into a Riemann BVP, and to obtain an integral form solution for π. We will not go more deeply into the analysis, which does not contain theoretical difficulty.

6.5 Case 5: $M_x = M_y = 0$

This is the limiting case where $x_2 = x_3 = 1$ and $y_2 = y_3 = 1$.

Lemma 6.5.1 *The branches X_0 and X_1 of the algebraic function X are, respectively, holomorphic and meromorphic in the complex plane $\mathbb{C}_y$ cuts along $[y_1 y_4]$. The image of the cut by $X(y)$ consists of two simple closed curves $\mathcal{M}_1$ and $\mathcal{M}_2$, with*

$$\mathcal{M}_1 = X[\overrightarrow{y_1, 1}], \qquad \mathcal{M}_2 = X[\overrightarrow{1, y_4}],$$

which are symmetrical with respect to the horizontal axis, and smooth (i.e. the direction of the tangent varies continuously), except at the point $x = 1$, which is a corner point if, and only if, the correlation coefficient r of the random walk in the interior of the quarter plane is not zero. The following situation holds:

- *If $r \neq 0$, then*

$$\mathcal{M}_1 \cap \mathcal{M}_2 = \{1\}, \quad \mathcal{M}_i \cap \Gamma = \{1\}, \quad i = 1, 2, \qquad (6.5.1)$$

$$\begin{cases} \mathcal{M}_1 \subset \overline{\mathcal{D}}, \\ \mathcal{M}_2 \subset \mathbb{C} - \mathcal{D}, \end{cases} \text{if } r < 0, \quad and \quad \begin{cases} \mathcal{M}_1 \subset \mathbb{C} - \mathcal{D}, \\ \mathcal{M}_2 \subset \overline{\mathcal{D}}, \end{cases} \text{if } r > 0.$$

$$(6.5.2)$$

In addition,

$$|X_0(y)| < 1 \quad \text{and} \quad |X_1(y)| > 1, \quad \forall y \in \Gamma - \{1\}.$$

- *If $r = 0$, then*

$$\mathcal{M}_1 \subset \overline{\mathcal{D}} \subset \mathcal{M}_2 \subset \mathbb{C} - \mathcal{D}.$$

Moreover the curves $\mathcal{M}_1, \mathcal{M}_2$, are tangent at $x = 1$. They can be identical if, and only if, the group is of order 4 (see condition (4.1.5)), in which case they coincide with the unit circle Γ. This is in particular always the case for the simple random walk.

In figure 6.5.1 the dotted curve is the unit circle, and one has drawn the contour $\mathcal{M}_1 \cup \mathcal{M}_2$, which has a self-intersection and is the image of the cut $[\overrightarrow{y_1 y_4}]$ by the mapping $y \to X(y)$, remembering that $X_0[\overrightarrow{y_1 y_4}] = \overline{X}_1[\overrightarrow{y_1 y_4}]$.

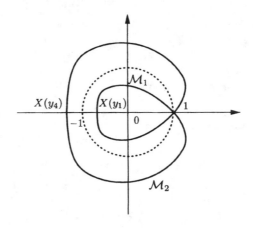

Fig. 6.5.1. The contour $\mathcal{M}_1 \cup \mathcal{M}_2$, for $r < 0$.

Similar properties hold for the algebraic function $Y(x)$, with the corresponding curves $\mathcal{L}_1$ and $\mathcal{L}_2$. ∎

Proof. To show (6.5.1), remark that if there exists $x \in \mathcal{M}_1 \cap \Gamma$, $x \neq 1$, then there exists θ, $\theta \in]0, 2\pi[$, such that the couples $(e^{\varepsilon i\theta}, Y_0(e^{\varepsilon i\theta}))$ and $(e^{\varepsilon i\theta}, Y_1(e^{\varepsilon i\theta}))$, $\varepsilon = \pm 1$, are solution of the fundamental equation $Q(x, y) = 0$. In addition, $Y_0(e^{i\theta}) \in [y_1, 1]$ and $Y_1(e^{i\theta}) > 1$, since $|Y_0(x)| < 1$, $\forall x \in \Gamma - \{1\}$.
Setting now $Y_j(e^{i\theta}) = z_j$, $j = 1, 2$, we obtain a system of equations, with respect to the variable x,

$$\begin{cases} a(x)z_1^2 + b(x)z_1 + c(x) = 0, \\ a(x)z_2^2 + b(x)z_2 + c(x) = 0. \end{cases}$$

But this system admits only the real solution $x = e^{i\theta}$, because all its coefficients are real. Thus $\theta = 0$ and (6.5.1) is proved.

As for (6.5.2), it suffices to analyze the ratio $\dfrac{\widetilde{c}(y)}{\widetilde{a}(y)}$, $y \in]1 - \varepsilon, 1[$, since on the cut $[y_1, 1]$, X_0 and X_1 are complex conjugate functions. We have

$$1 - \frac{\widetilde{c}(y)}{\widetilde{a}(y)} = \frac{(\widetilde{a}'(1) - \widetilde{c}'(1))(y - 1) + \mathcal{O}(y - 1)^2}{\widetilde{a}(y)}.$$

Now, using $M_x = 0$ and letting R denote the covariance of the random walk in the interior of the quarter-plane, one checks easily the following equivalence

$$R \overset{\text{def}}{=} \widetilde{a}'(1) - \widetilde{c}'(1) = p_{11} + p_{-1,-1} - p_{-11} - p_{1,-1} \Longleftrightarrow r < 0,$$

which corresponds exactly to (6.5.2).

The claim of the lemma that $\mathcal{M}_1$ has a corner point at $x = 1$ can be shown by differentiating twice with respect to y at $y = 1$ the equation

$$\widetilde{a}(y) X^2(y) + \widetilde{b}(y) X(y) + \widetilde{c}(y) = 0.$$

After an easy manipulation, it appears that the derivative

$$t \overset{\text{def}}{=} \frac{dX_0(y)}{dy} \bigg|_{y=1}$$

of the branch X_0, in the cut plane $\mathbb{C} - [y_1, y_2]$, is one of the roots of the second degree equation

$$\widetilde{a}(1) t^2 + Rt + a(1) = 0. \tag{6.5.3}$$

Since $M_x = M_y = 0$, we also have

$$b(1) = -2a(1), \quad \widetilde{b}(1) = -2\widetilde{a}(1).$$

Hence, the discriminant Δ of equation (6.5.3) is given by

$$\Delta = R^2 - 4a(1)\widetilde{a}(1) = R^2 - b(1)\widetilde{b}(1).$$

But $|R| < \min(|b(1)|, |\widetilde{b}(1)|)$, so that Δ is negative and the roots of (6.5.3) are finite and complex conjugate. Thus $\mathcal{M}_1$ has a corner point if, and only if, $r \neq 0$. In a similar way, the derivative $s \overset{\text{def}}{=} \dfrac{dY_0(x)}{dx} \bigg|_{x=1}$ satisfies the equation

$$a(1) s^2 + Rs + \widetilde{a}(1) = 0.$$

The assertions of the lemma concerning the case $r = 0$ and the group of order 4 can easily be achieved by continuity arguments with respect to the parameters, like in chapter 5. The proof of lemma 6.5.1 is concluded. ∎

Theorem 6.5.2

(i) *If $r \leq 0$, then π [resp. $\widetilde{\pi}$] is a holomorphic solution of the Riemann-Carleman problem*

$$\pi(t)A(t) - \pi(\alpha(t))A(\alpha(t)) = g(t), \quad t \in \mathcal{M}_1, \qquad (6.5.4)$$

where the functions A and g are defined in (5.1.6).

(ii) *If $r > 0$, then π [resp. $\widetilde{\pi}$] can be continued as a meromorphic function in the region $G_{\mathcal{M}_1}$ [resp. $G_{\mathcal{L}_1}$]. The poles of π in $G_{\mathcal{M}_1} - \mathcal{D}$ [resp. $\widetilde{\pi}$ in $G_{\mathcal{L}_1} - \mathcal{D}$] are the eventual zeros of q [resp. $\widetilde{q}$].*

(iii) *The function π and $\widetilde{\pi}$ cannot be analytically continued to a region encompassing the point 1, and hence the conditions of lemma 2.2.1 are not satisfied.* ∎

Proof.   One proceeds exactly as in chapter 5, eliminating $\widetilde{\pi}$ which must be continuous on the two edges of the slit $[y_1, 1]$. This yields (6.5.4), exactly as was obtained (5.1.5), and the point (iii) is clear. The technical novelty is that the BVP (6.5.4) belongs to the class of generalized problems (see [35, 61]), since the contour $\mathcal{M}_1$ has a corner point. We will return to this fact later, when computing the index. ∎

From now on, one will assume $r < 0$. The reader will convince himself that the forthcoming proofs can easily be transposed to the case $r > 0$, up to some technicalities arising from possible poles in the domain $G_{\mathcal{M}}$, which then contains the unit circle as shown in (6.5.2). Moreover, probabilistic arguments show that the random walk is never ergodic for $r > 0$. In fact, the mean first entrance time of the random walk into the axes, when starting from some arbitrary point with strictly positive coordinates, is infinite for any $r > 0$ (see [29]).

As in theorem 5.4.1 for the case of genus 1, we will derive the conditions for the functions π and $\widetilde{\pi}$ to be analytic in the open disc $\mathcal{D}$ and continuous on its boundary Γ. The argument will mimic those of section 5.4, and all the *drifts* coming into the computations have been defined in (1.2.3).

Lemma 6.5.3 *Let*

$$\begin{cases} \lambda_x \stackrel{def}{=} \sum_{ij} i^2 p_{ij} = 2\widetilde{a}(1), \quad \lambda_y \stackrel{def}{=} \sum_{ij} j^2 p_{ij} = 2a(1), \\[2mm] \sigma \stackrel{def}{=} \dfrac{dq(x, Y_0(x))}{dx}\Big|_{x=1} = M'_x - \dfrac{RM'_y}{\lambda_y} - i\dfrac{M'_y\sqrt{|\Delta|}}{\lambda_y} \stackrel{def}{=} \rho e^{i\theta}, \\[3mm] \widetilde{\sigma} \stackrel{def}{=} \dfrac{d\widetilde{q}(X_0(y), y)}{dy}\Big|_{y=1} = M''_y - \dfrac{RM''_x}{\lambda_x} - i\dfrac{M''_x\sqrt{|\Delta|}}{\lambda_x} \stackrel{def}{=} \widetilde{\rho}e^{i\widetilde{\theta}}. \end{cases} \qquad (6.5.5)$$

The BVP corresponding to (6.5.4) has a reduced form corresponding to (5.4.19), the index of which is given by

$$\chi = -L - M + \left| \frac{\theta + \widetilde{\theta} + \arccos(-r)}{\pi} - 1 \right|. \tag{6.5.6}$$

Moreover, π and $\widetilde{\pi}$ correspond to probabilistic distributions if, and only if,

$$\left| \frac{\theta + \widetilde{\theta} + \arccos(-r)}{\pi} - 1 \right| = 1 \tag{6.5.7}$$

∎

Proof.   We follow the method and the notation proposed in lemma 5.4.2. The main discrepancy with respect to chapter 5 comes from the discontinuity of the derivative of $X_0(y)$ [resp. $Y_0(x)$] at $y = 1$ [resp. $x = 1$], which renders slightly sharper the calculation of the *reduced index*. Following (5.4.5), the basic step is to estimate the quantity

$$\chi_1 \stackrel{\text{def}}{=} -\mathcal{I}nd[A(t)]_{\mathcal{M}_1} = \zeta + \widetilde{\zeta},$$

with

$$\zeta \stackrel{\text{def}}{=} -\mathcal{I}nd[q(x, Y_0(x))]_{\mathcal{M}_1}, \qquad \widetilde{\zeta} \stackrel{\text{def}}{=} \mathcal{I}nd[\widetilde{q}(X_0(y), y)]_{\overline{[y_1,1]}}.$$

Letting now

$$\begin{cases} \delta = \mathcal{I}nd[q(x, Y_0(x))]_{\Gamma}, \\ \widetilde{\delta} = \mathcal{I}nd[\widetilde{q}(X_0(y), y)]_{\Gamma}, \end{cases} \tag{6.5.8}$$

we can write

$$\begin{cases} \zeta = N_Z[q(x, Y_0(x)), \mathcal{D}] - N_Z[q(x, Y_0(x)), G_{\mathcal{M}_1}] - L - \delta + 1 - \dfrac{\alpha}{\pi} \\ \widetilde{\zeta} = N_Z[\widetilde{q}(X_0(y), y), \mathcal{D}] - M - \widetilde{\delta} + \dfrac{1}{2}, \end{cases} \tag{6.5.9}$$

where $0 \leq \alpha \leq \pi$ denotes the angle between the two tangents of the curves $\mathcal{M}_1$ and Γ at the point 1. The derivation of (6.5.9) deserves some explanation. We consider the domain between the curve $\mathcal{M}_1$ and Γ. Then the famous principle of the argument for the functions q, on the boundary of this domain, does not apply without precaution, since q is discontinuous at the a corner point 1. Indeed the variation has to be taken as a Stieltjes integral in principal value, and requires the generalized form of the Sokhotski-Plemelj formulae (see section 5.2.2 and e.g. [61]):

$$\begin{cases} \Phi^+(t) = \left(1 - \dfrac{\alpha}{2\pi}\right)\varphi(t) + \dfrac{1}{2i\pi}\displaystyle\int_{\mathcal{L}} \dfrac{\varphi(s)ds}{s-t}, \\ \Phi^-(t) = -\dfrac{\alpha}{2\pi}\varphi(t) + \dfrac{1}{2i\pi}\displaystyle\int_{\mathcal{L}} \dfrac{\varphi(s)ds}{s-t}, \end{cases}$$

where $0 \leq \alpha \leq 2\pi$ is the angle between the two half-tangents at the corner point. We apply these remarks to calculate δ and $\widetilde{\delta}$ defined by (6.5.8).

Choosing the argument of an arbitrary complex number to take its values on the interval $[0, 2\pi[$, we get

$$\begin{cases} \delta = 1 - \dfrac{\theta}{\pi}, \\ \widetilde{\delta} = 1 - \dfrac{\widetilde{\theta}}{\pi}, \end{cases}$$

whence

$$\delta + \widetilde{\delta} = 2 - \frac{\theta + \widetilde{\theta}}{\pi}.$$

For the last point of the lemma concerning probabilistic solutions, we use the result derived in section 5.4 (which still holds, provided the index be properly defined, since $\mathcal{M}_1$ is not a smooth Lyapounov contour, see e.g. [35]). Hence, for π and $\widetilde{\pi}$ to be proper probability distributions, it is necessary and sufficient to have

$$\chi = -L - M + 1,$$

which reduces easily to (6.5.7). The proof of the lemma is terminated. ■

Remark 6.5.4 *The* ergodicity conditions *for the two-dimensional random walk, obtained in a purely probabilistic way in [29]), write as follows:*

$$M'_x < 0, \quad M''_y < 0, \quad and \quad \lambda_x \frac{M'_y}{M'_x} + \lambda_y \frac{M''_x}{M''_y} > 2R.$$

The reader can check that they do not coincide with (6.5.7) and bring to light a spectacular phenomenon. We have seen in theorem 6.5.2 the impossibility of using lemma 2.2.1. Indeed, $\pi(x, y)$ is analytic in the open domain $\mathcal{D} \times \mathcal{D}$, but the continuity on the boundary $\Gamma \times \Gamma$ need not hold, unless additional conditions on the parameters are in force. This was recently observed by probabilistic arguments in [2].

Essentially because of technicalities concerning the *explicit* expression of conformal mappings, it will be more convenient to pose the BVP (6.5.4) on the curve

$$\widetilde{\mathcal{M}} \stackrel{\text{def}}{=} \mathcal{M}_1 \cup [\overline{x_1, 1}]$$

shown in figure 6.5.2.

Lemma 6.5.5 *The simply-connected domain $G_{\widetilde{\mathcal{M}}}$ is conformally mapped by one branch of the function $t \to x^{-1}(t)$ onto the lens-shaped domain $\mathcal{H}$ bounded by two circular arcs, in the Z-plane as shown in figure 6.5.2. In addition, $\mathcal{H}$ is conformally mapped onto the upper half-plane by the function*

$$w(z) = \left(\frac{z - z_0}{z_0 z - 1} \right)^{\frac{\pi}{\psi}}, \tag{6.5.10}$$

where ψ is the angle between the two arcs. ■

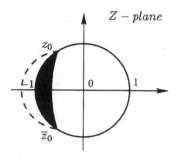

Fig. 6.5.2. The conformal transform of $\widetilde{\mathcal{M}}$

Proof.   The algebraic curve corresponding to $Q(x, y) = 0$ admits a rational uniformization, of the following general form (see section 6.4), $\forall z \in \mathbb{C}$,

$$\begin{cases} x(z) & = \dfrac{x_4 + x_1}{2} + \dfrac{x_4 - x_1}{4}\left(z + \dfrac{1}{z}\right), \\[3mm] y(u) & = \dfrac{y_4 + y_1}{2} + \dfrac{y_4 - y_1}{4}\left(\eta(z) + \dfrac{1}{\eta(z)}\right). \end{cases} \qquad (6.5.11)$$

The curve $\mathcal{M}_1$ corresponds in the z-plane to two circular arcs, drawn in figure 6.5.2, which are inverse with respect to the unit circle Γ, so that one of them is located inside $\mathcal{D}$. They cross Γ at the conjugate points $z_0, \overline{z}_0$, where

$$x(z_0) = x(\overline{z}_0) = 1.$$

The function in (6.5.11) is standard (see e.g [71]) and the proof of the lemma is terminated.   ∎

Since the mapping $w \to w + \dfrac{1}{w}$ maps the unit disk onto the plane cut along the slit $[-1, +1]$, the function π is finally obtained from the integral formula (5.4.21), using (6.5.6) for the index, and (6.5.10).

7. Miscellanea

In this book we have described completely only some topics related to the subject. There are however a lot of directly related questions not included into the present book.

7.1 About Explicit Solutions

Besides considering BVP in the complex plane, there are different alternatives. The first one is to consider a BVP on a Riemann surface, the second one to work on the universal covering, as shown in section 5.5. The direct passage to the universal covering itself, by uniformization, involves Weierstrass functions. In [46], the functions π and $\widetilde{\pi}$ are represented in terms of convergent series (as for meromorphic functions) and convergent products (as for entire functions). In all these approaches, one must localize the zeros and the poles of q and $\widetilde{q}$ in some regions. Even if this can be easily done for some parameter values, the technical details are tedious in the general case.

The method of solution presented in the book can be applied to solve non stationary problems, by means of an intermediate Laplace transform [27].

In [3], one has considered a diffusion process in $\mathbb{Z}_+^2$. The determination of its invariant measure is equivalent to solve a BVP for an elliptic operator on an hyperbola (!), allowing for explicit solutions. Here the generating functions are replaced by Laplace-transforms and the unit disk by the right half-plane.

In other problems, especially in many models encountered in queueing theory, jumps inside the quarter plane are not bounded, and it is no more possible to speak of algebraic curve. Nonetheless, it is still possible to associate the BVP approach in the complex plane with conformal mapping techniques (or Fredholm equations). In this connection, the reader can see [19, 4].

7.2 Asymptotics

Under this title, several important areas are in fact concerned.

7.2.1 Large Deviations and Stationary Probabilities

Even if asymptotic problems were not mentioned in this book, they have many applications and are mostly interesting for higher dimensions. We recall here the results of [55] concerning the asymptotic behavior of the stationary probabilities. We shall consider the *simple random walk* defined in chapter 2, i.e. inside the quarter plane all the transition probabilities are zero, but $p_{01}, p_{10}, p_{-1,0}, p_{0,-1}$. In addition the following assumptions are made on the drift vectors $\mathbf{M}, \mathbf{M}', \mathbf{M}''$ defined in (1.2.3) :

$$\mathbf{M} < \mathbf{0}, \quad M'_y \neq 0, \quad M''_x \neq 0. \tag{7.2.1}$$

The choice of the simple random walk is not crucial for the applicability of the analytical methods, but it simplifies considerably the computations. Also, the case where only one component of $\mathbf{M}$ in (7.2.1) is negative, can be analyzed via similar methods.

Some new facts about the Riemann surface will be needed. Let $x(s), y(s)$ be meromorphic functions on $\mathbf{S}$ defining the coverings of the x-plane and y -plane respectively.

1. We know that the algebraic function $Y(x)$ [resp. $X(y)$] has exactly four branch points x_i [resp. y_i], where

$$\begin{cases} 0 < x_1 < x_2 < 1 < x_3 < x_4, \\ 0 < y_1 < y_2 < 1 < y_3 < y_4. \end{cases}$$

2. Let $\mathbf{S}_r = \{s : x(s) \in \mathbb{R}, y(s) \in \mathbb{R}\}$ be the set of all real points of $\mathbf{S}$. Then $\mathbf{S}_r$ consists of two disjoint analytic closed curves, homologous to one of the elements of the normal homology basis on $\mathbf{S}$, more exactly to the one different from $x^{-1}(\{x : |x| = 1\})$. These curves will be denoted by F_0, F_1, with F_0 satisfying

$$x_2 \leq x(s) \leq x_3 \quad \text{and} \quad y_2 \leq y(s) \leq y_3, \ \forall s \in F_0.$$

F_0 contains an ordered set of eight characteristic points $s_0, \ldots, s_7$, which are defined as follows:

$$\begin{cases} s_0 = (1, 1), \ s_1 = \left(\sqrt{\dfrac{p_{0,-1}}{p_{01}}}, y_2\right), \ s_2 = \left(\dfrac{p_{0,-1}}{p_{01}}, 1\right), \\[3mm] s_3 = \left(x_3, \sqrt{\dfrac{p_{-1,0}}{p_{10}}}\right), \ s_4 = \left(\dfrac{p_{0,-1}}{p_{01}}, \dfrac{p_{-1,0}}{p_{10}}\right), \ s_5 = \left(\sqrt{\dfrac{p_{0,-1}}{p_{01}}}, y_3\right), \\[3mm] s_6 = \left(1, \dfrac{p_{-1,0}}{p_{10}}\right), \ s_7 = \left(x_2, \sqrt{\dfrac{p_{-1,0}}{p_{10}}}\right). \end{cases}$$

3. The function $\phi_\gamma(s) = |xy^\gamma|$, $0 \leq \gamma \leq 1$, has in the set $\{x \neq 0, y \neq 0\}$ four non-degenerate critical points $s_i(\gamma), i = 1, \ldots, 4$, which are defined by the two of equations

$$Q(x, y) = 0, \quad y\frac{\partial}{\partial y}Q(x, y) = \gamma x\frac{\partial}{\partial x}Q(x, y).$$

Each $s_i(\gamma)$ is uniquely defined and depends continuously on γ, remarking that $x(s_i(0)) = x_i$.

Moreover, $s_2(\gamma), s_3(\gamma) \in F_0$ and $x_i(\gamma) = x(s_i(\gamma)), y_i(\gamma) = y(s_i(\gamma))$ are real. For $\gamma = 1$, one can put $s_1(1) = 0, s_4(1) = \infty$, and hence, for the critical points $s_2(1), s_3(1)$, the above assertions hold. We have also

$$\begin{cases} 1 < x_3(1) < x_3(\gamma) < x_3(0) = x_3, \\ y_3(0) = \sqrt{\dfrac{p_{0,-1}}{p_{01}}} < y_3(\gamma) < y_3(1). \end{cases}$$

It appears that the asymptotics of the stationary probabilities is determined either by the critical point $s_3(\gamma)$ or by the zeros of q_ξ or $\tilde{q}_\eta$. Note as a reminder that ξ and η are the Galois automorphisms on $\mathbf{S}$ constructed in section 2.4, and such that

$$\xi(x, y) = \left(x, \frac{p_{0,-1}}{p_{01}y}\right), \quad \eta(x, y) = \left(\frac{p_{-1,0}}{p_{10}x}, y\right).$$

Now in the parameter space $\mathcal{P} \times \{\gamma : 0 < \gamma \leq 1\}$, with $\mathcal{P}$ given in section 2.3.1, we introduce the subsets

$$\begin{cases} \mathcal{P}_{--} = \left\{(p, \gamma) : q\left(x_3(\gamma), \dfrac{p_{0,-1}}{p_{01}y_3(\gamma)}\right) \leq 0, \ \tilde{q}\left(\dfrac{p_{-1,0}}{p_{10}x_3(\gamma)}, y_3(\gamma)\right) \leq 0\right\}, \\ \mathcal{P}_{+-} = \left\{(p, \gamma) : q\left(x_3(\gamma), \dfrac{p_{0,-1}}{p_{01}y_3(\gamma)}\right) > 0, \ \tilde{q}\left(\dfrac{p_{-1,0}}{p_{10}x_3(\gamma)}, y_3(\gamma)\right) \leq 0\right\}, \end{cases}$$

and $\mathcal{P}_{-+}, \mathcal{P}_{++}$ accordingly. The main result is given by the next theorem.

Theorem 7.1 *Let $m, n \to \infty$ so that $\dfrac{n}{m} \to \gamma, 0 < \gamma \leq 1$. Then we have*

$$\pi(m, n) \sim \begin{cases} \dfrac{C_1}{\sqrt{m}}x_3^{-m}(\gamma)y_3^{-n}(\gamma) & \text{in } \mathcal{P}_{--}, \\ C_2 x_0^{-m}(\gamma)y_0^{-n}(\gamma) & \text{in } \mathcal{P}_{-+}, \\ C_3 x_5^{-m}(\gamma)y_5^{-n}(\gamma) & \text{in } \mathcal{P}_{+-}, \\ C_4 x_0^{-m}(\gamma)y_0^{-n}(\gamma) + C_5 x_5^{-m}(\gamma)y_5^{-n}(\gamma) & \text{in } \mathcal{P}_{++}, \end{cases} \tag{7.2.2}$$

where the C_i's are constants (depending on γ) and

$$\begin{cases} 1 < x_0(\gamma) < x_3(\gamma), \quad 1 < y_0(\gamma) < \dfrac{p_{0,-1}}{p_{01}y_3(\gamma)}, \\ 1 < x_5(\gamma) < \dfrac{p_{-1,0}}{p_{10}x_3(\gamma)}, \quad 1 < y_5(\gamma) < y_3(\gamma) \end{cases}$$

are solutions of the respective systems

$$\begin{cases} Q(x,y) = 0, & q(x, \xi y) = 0, \quad \text{for } x_0(\gamma), y_0(\gamma); \\ Q(x,y) = 0, & \widetilde{q}(\eta x, y) = 0, \quad \text{for } x_5(\gamma), y_5(\gamma). \end{cases}$$

For $\gamma = 0$, the asymptotic behavior could be obtained (although this was not done) by means of the method proposed in [66]. ∎

We will only give a brief sketch of the methods used to prove the main result (7.2.2). The stationary probabilities can be represented by two-dimensional Cauchy integrals, which, using Leray residues (see e.g. [1]), can be reduced to one-dimensional integrals over some cycle on the Riemann surface. The asymptotics will be thus defined either by the steepest descent point (*saddle-point*) or by a pole, that could be encountered while moving the integration contour on the surface. Both the saddle point and the possible pole belong to the real part of the algebraic curve.

Martin Boundary Roughly, let us simply say that the so-called *Martin boundary* (MB) describes the way in which a process escapes to infinity (see e.g. [69]). For non-smooth regions, probabilistic methods rarely yield a complete characterization, while complex analysis appears promising. To calculate the MB, one needs the asymptotics of Green functions, and thus of the MB kernel. This requires a little bit more effort than large deviation asymptotics. Transient random walks in $\mathbb{Z}_+^2$, with the same kind of jumps as in the present book, were considered in [42]. Starting from convenient functional equations, it was possible by analytic continuation arguments to obtain the first singularity and the saddle-point (which form the main contribution to the MB), but still not a complete explicit expression.

7.3 Generalized Problems and Analytic Continuation

We have already seen how analytic continuation is crucial in most of the problems, including asymptotics. Undoubtedly the first step toward a generalization, in the case of jumps bounded in modulus by a finite number n, is the analytic continuation process. Here there are $2n$ unknown functions, $\pi_i(s), \widetilde{\pi}_i(s)$, which must be analytic in the connected domain $\mathcal{E} \subset \mathbf{S}$,

$$\mathcal{E} = \{|x(s)| < 1, |y(s)| < 1\}.$$

Then a functional equation can be obtained, on a Riemann surface $\mathbf{S}$ of arbitrary genus, which has the form

$$\sum_{i=1}^{n} \Big(q_i(s)\pi_i(x(s)) + \widetilde{q}_i(s)\widetilde{\pi}_i(y(s)) \Big) + q_0(s) = 0 \tag{7.3.1}$$

where $q_i(s), \widetilde{q}_i(s)$ are meromorphic on $\mathbf{S}$. The next results where proved in [56].

1 $\pi_i(s), \widetilde{\pi}_i(s)$ can be continued "infinitely" outside $\mathcal{E}$ and can have only algebraic branch points in the complex plane. The resulting covering surface is isomorphic to the disk but it is not the universal covering of **S**. The reasons become clear when we consider the sequence of fields

$$F_0 \subset F_1 \subset F_2 \subset \dots , \tag{7.3.2}$$

where

- F_0 is the field of meromorphic functions on **S**;
- F_{2i} is the minimal Galois extension of $C(y)$, containing $F_{2i-1}, i \geq 1$;
- F_{2i+1} is the minimal Galois extension of $C(x)$ containing $F_{2i}, i \geq 0$.

In the generic situation $\pi_i(s), \widetilde{\pi}_i(s)$ can only be considered as meromorphic functions on the limit of $\mathbf{S}_i$-Riemann surface of F_i, or, more exactly, as sections of the inductive limit $\lim_{i \to \infty}$ of the bundles of meromorphic functions on $\mathbf{S}_i$;

- there is a necessary and sufficient condition for the finiteness of the sequence F_i, showing that stabilization rarely occurs. In case of the stabilization they are meromorphic on S_k for some finite k;

- if the problem can be solved by Wiener-Hopf factorization techniques, then the chain stops at F_2.

In the class of generalized problems, one finds also the famous queueing model *Joining the shorter queue*, which was solved in the non-symmetrical case in [22, 39]. Here there is no space homogeneity: the quarter plane is separated into two homogeneous regions, and one must write a functional equation for each region. This gives rise to the following system:

$$\begin{cases} Q_1(x,y)\pi_1(x,y) = q_{11}(x,y)\pi_1(x) + q_{12}(x,y)\pi_2(x) + \widetilde{q}_1(x,y)\widetilde{\pi}_1(y) , \\ Q_2(x,y)\pi_2(x,y) = q_{21}(x,y)\pi_1(x) + q_{22}(x,y)\pi_2(x) + \widetilde{q}_2(x,y)\widetilde{\pi}_2(y) . \end{cases}$$

There are 4 unknown functions of one variables. The algebraic curves corresponding to Q_1 and Q_2 are of genus zero. It is possible to write, in some region, a non-Nœtherian BVP (i.e. its index is not finite). Upon combining Q_1 and Q_2, let us say simply that we are in the situation of non-Galois extension. It was proved that all functions can be analytically continued as meromorphic functions to the whole complex plane. The symmetrical case, i.e. $Q_1 = Q_2$, reduces to a single functional equation on a curve of genus zero, and it was originally solved in [32].

Finally, it is not necessary to insist on the usefulness of getting results for random walks if $\mathbb{Z}_+^N, N > 2$. For a first step in this direction, see [65]. The general problem of unique continuation for solutions of functional equations on spaces with group actions was considered in [58].

7.4 Outside Probability

Operator Factorization and C^*-Algebras There was a great activity concerning the *index problem* for pseudo-differential operators in domains with a smooth boundary, and one of the central results is the so-called Atya-Zinger index theorem. In the present book, we have dealt with the simplest case of non-smooth boundary (one corner), and the derivations of the ergodicity conditions, via the index, presented in sections 5.4 6.5 are new.

In larger dimensions, the index problem, especially for Toeplitz operators in $\mathbb{Z}_+^N$, Wiener-Hopf equations in $\mathbb{R}_+^N$ and BVP for N complex variables, is still largely open. The reason resides in the inductiveness property: indeed, dimension N demands much finer properties for the related problems in dimension $N - 1$ [59]. However, formal inductive solution for general N can be given in terms of C^*-algebras and by factorization of functions on the circle, taking their values in the set of compact operators in Hilbert spaces (see [57]).

Physics Some relationships with the so-called *integrable models* in statistical physics are described in [59], but they seem to be much deeper.

References

[1] L.A Aizenberg and A.P. Yuzhakov. *Integral Representations and Residues in Multidimensional Complex Analysis*, volume 58 of *Translation of Mathematical Monographs*. AMS, 1983.

[2] S. Aspandiiarov and R. Iasnogorodski. General results on stationary distributions for countable Markov chains and their applications. *Bernoulli*, 1999. To appear.

[3] F. Baccelli and G. Fayolle. Analysis of models reducible to a class of diffusion processes in the positive quarter plane. *SIAM J. Appl. Math.*, 47(6):1367–1386, Decembre 1987.

[4] D.E. Barrett, H.R. Gail, S.L. Hantler, and B.A. Taylor. Varieties in a two dimensional polydisk with univalent projection at the boundary. Technical Report 15848, IBM, Research Division, T.J. Watson Research Center, Yorktown Heights, NY 10598, 1990.

[5] H. Bateman. *Higher Transcendental Functions*. McGraw-Hill, 1953. H. Bateman manuscript project; A. Erdelyi editor.

[6] J.P.C. Blanc. Application to the theory of boundary value problems in the analysis of a queueing model with paired services. Mathematical Centre Tracts 153, Mathematisch Centrum, Amsterdam, 1982.

[7] J.P.C. Blanc. On the relaxation times in open queueing networks. In O.J. Boxma and R. Syski, editors, *Queueing Theory and its Applications. Liber Amicorum for J.W. Cohen*, volume 7 of *CWI Monographs*, pages 235–259. North-Holland, 1988.

[8] J.P.C. Blanc, R. Iasnogorodski, and Ph. Nain. Analysis of the $M/GI/1 \rightarrow ./M/1$ queueing model. *Queueing Systems*, 3(2):129–156, April 1988.

[9] S. Bochner and W.T. Martin. *Several Complex Variables*. Princeton University Press, 1948.

[10] O.J. Boxma and W.P. Groenendijk. Two queues with alternating service and switching times. In O.J. Boxma and R. Syski, editors, *Queueing Theory and its Applications. Liber Amicorum for J.W. Cohen*, pages 261–282. North Holland, 1989.

[11] K.L. Chung. *Markov Chains with Stationary Transition Probabilities*. Springer Verlag, second edition, 1967.

[12] E.G. Coffman, G. Fayolle, and I. Mitrani. Sojourn times in a tandem queue with overtaking: reduction to a boundary value problem. *Stochastic Models*, 2(1):43–65, 1986.

[13] E.G. Coffman Jr., G. Fayolle, and I. Mitrani. Two queues with alternating service periods. In P.J. Courtois and G. Latouche, editors, *Performance'87*, pages 227–237. North-Holland, 1983.

152   References

[14] J.W. Cohen. On the M/G/2 queueing model. *Stochastic Processes and their Applications*, 12:231–248, 1982.

[15] J.W. Cohen. On the analysis two-dimensional queueing problems. In G. Iazeolla, P.J. Courtois, and A. Hordijk, editors, *Mathematical Computer Performance and Reliability*, pages 17–32. North-Holland, 1984.

[16] J.W. Cohen. A two-queue model with semi-exhaustive alternating service. In P.J. Courtois and G. Latouche, editors, *Performance'87*, pages 19–37. North-Holland, 1987.

[17] J.W. Cohen. A two-queue, one-server model with priority for the longer queue. *Queueing Systems*, 2:261–283, 1987.

[18] J.W. Cohen. Boundary value problems in queueing theory. *Queueing Systems*, 3:97–128, 1988.

[19] J.W. Cohen and O.J. Boxma. *Boundary Value Problems in Queueing System Analysis*. North-Holland, 1983.

[20] S.J. de Klein. Fredholm integral equations in queueing analysis. Master's thesis, University of Utrecht, 1988.

[21] M. Eisenberg. Two queues with alternating service. *SIAM Journal of Applied Mathematics*, 36:287–303, 1979.

[22] G. Fayolle. *Méthodes analytiques pour les files d'attente couplées*. Doctorat d'État ès Sciences Mathématiques, Université Paris VI, Novembre 1979.

[23] G. Fayolle. On functional equations of one and two complex variables arising in the analysis of stochastic models. In A. Hordijk G. Iazeolla, P.J. Courtois, editor, *Mathematical Computer Performance and Reliability*, pages 55–75. North-Holland, 1984.

[24] G. Fayolle. A functional approach to the problem of sojourn times in simple queueing networks with overtaking. In Luis F.M. de Moraes, E. de Souza e Silva, and simple queueing networks with overtaking, editors, *Data Communication Systems and their Performance*. Editora Campus, June 1987.

[25] G. Fayolle and M.A. Brun. On a system with impatience and repeated calls. In O.J. Boxma and R. Syski, editors, *Queueing Theory and its Applications. Liber Amicorum for J.W. Cohen*, volume 7 of *CWI Monographs*, pages 283–305. North-Holland, 1988.

[26] G. Fayolle and R. Iasnogorodski. Two coupled processors: the reduction to a Riemann-Hilbert problem. *Zeitschrift für Wahrscheinlichkeitstheorie und Verwandte Gebiete*, 47:325–351, 1979.

[27] G. Fayolle, R. Iasnogorodski, and I. Mitrani. The distribution of sojourn times in a queueing network with overtaking: Reduction to a boundary value problem. In S.K. Tripathi A.K Agrawala, editor, *Proceedings of the 9th International. Symp. on Comp. Perf. Modelling, Measurement and Evaluation*, pages 477–486. North-Holland, 1983.

[28] G. Fayolle, P.J.B. King, and I. Mitrani. The solution of certain two-dimensional markov models. *Advances in Applied Probability*, 14:295–308, 1982.

[29] G. Fayolle, V.A. Malyshev, and M.V. Menshikov. *Topics in the Constructive Theory of Countable Markov Chains*. Cambridge University Press, 1995.

[30] L. Flatto. Two parallel queues created by arrivals with two demands - II. *SIAM Journal of Applied Mathematics*, 45(5):861–878, 1985.

[31] L. Flatto and S. Hahn. Two parallel queues created by arrivals with two demands - I. *SIAM Journal of Applied Mathematics*, 44(5):1041–1053, 1984.

[32] L. Flatto and H.P. McKean. Two queues in parallel. *Communications on Pure and Applied Mathematics*, 30:255–263, 1977.

[33] L.R Ford. *Automorphic Functions*. Chelsea, New-York, N.Y., second edition, 1972. Reprinted.

[34] O. Forster. *Lectures on Riemann surfaces*. Springer Verlag, 1981.

[35] F.D. Gakhov. *Boundary value problems*. Pergamon Press, 1966.

[36] Y. Gromack and V.A. Malyshev. Probability of hitting a finite set for a random walk in a quarter plane with absorbing boundary. In *Thesis of International Conference in Probability*, volume 1, Vilnius, 1973.

[37] R. Hartshorne. *Algebraic Geometry*. Springer Verlag, 1977.

[38] A. Hurwitz and R. Courant. *Funktionen Theorie*. Springer Verlag, fourth edition, 1964.

[39] R. Iasnogorodski. *Problèmes frontières dans les files d'attente*. Doctorat d'État ès Sciences Mathématiques, Université Paris VI, Novembre 1979.

[40] F.P. Kelly. *Reversibility and Stochastic Networks*. Wiley, London, 1979.

[41] K. Kendig. *Elementary Algebraic Geometry*. Springer Verlag, 1977.

[42] I. Kurkova and V.A. Malyshev. Martin boundary and elliptic curves. *Markov Processes and Related Fields*, 4(2), 1998.

[43] S. Lang. *Algebra*. Addison-Wesley, second edition, 1984.

[44] G. S Litvintchuk. *Boundary value problems and singular integral equations with shift*. Nauka Editions, 1977. In Russian.

[45] V.A. Malyshev. The solution of the discrete Wiener-Hopf equations in a quarter-plane. *Dockl. Akad. Nauk USSR*, 187:1243–1246, 1969.

[46] V.A. Malyshev. *Random walks. The Wiener-Hopf equations in a quadrant of the plane. Galois automorphisms*. Moscow State University Press, 1970.

[47] V.A. Malyshev. Positive random walks and Galois theory. *Uspehi Matem. Nauk*, 1:227–228, 1971.

[48] V.A. Malyshev. Positive random walks and generalised elliptic integrals. *Dokl. Akad. Nauk USSR*, 196(3):516–519, 1971.

[49] V.A. Malyshev. Wiener-Hopf equations in a quadrant of the plane, discrete groups and automorphic functions. *Math. USSR Sbornik*, 13:491–516, 1971.

[50] V.A. Malyshev. Analytical method in the theory of positive two-dimensional random walks. *Siberian Math. Journal*, 6:1314–1329, 1972.

[51] V.A. Malyshev. Non-standard Markov queues. In *Queueing theory*, pages 91–96. Second Soviet Symp. in Queueing Theory, 1972.

[52] V.A. Malyshev. Simple explicit formulas for some random walks in a quarter plane. In A.N. Kolmogorov, editor, *Veroyatnostnye Metody Issledovania*, volume 41, pages 14–22. Moscow State University, 1972.

[53] V.A. Malyshev. Stationary distribution for degenerate random walks in a quarter plane. *Vestnik Moskov Univ. Sr. 1 Mat. Meh.*, 2:18–24, 1972. with Y. Gromack.

[54] V.A. Malyshev. Analytic continuation in boundary value problems for two complex variables. *Functional Analysis and Applications*, 7(3):85–87, 1973.

[55] V.A. Malyshev. Asymptotic behaviour of stationary probabilities for two dimensional positive random walks. *Siberian Mathem. Journal*, 14(1):156–169, 1973.

[56] V.A. Malyshev. *Boundary value problems for two complex variables and applications*. Thesis for the degree of doctor in mathematics, Moscow University, 1973.

[57] V.A. Malyshev. Factorisation of functions with values in algebras of multidimensional Toeplitz operators. *Uspehi Matem. Nauk.*, 28(2):237–238, 1973.

[58] V.A. Malyshev. Linear recurrent equations on G-spaces. Moscow, MGU, 1974.

[59] V.A. Malyshev. Wiener-Hopf equations and their applications in probability theory. *Probability theory. Mathematical Statistics. Theoretical Cybernetics Itogi Nauki i Techniki*, 13:5–35, 1976. English translation in Journal of Soviet Maths., Vol. 7 (1977), No 2, pp.129–148.

[60] N. Mikou. *Modèles de réseaux de files d'attente avec pannes*. Doctorat d'État ès Sciences Mathématiques, Université Paris-Sud, Décembre 1981.

[61] N.I. Muskhelishvili. *Singular Integral Equations*. P. Noordhoff, 1953.

[62] Ph. Nain. On a generalization of the preemptive resume priority. *Advances in Applied Probability*, 18:255–273, 1986.

[63] Ph. Nain. *Applications des méthodes analytiques à la modélisation des systèmes informatiques*. Doctorat d'État ès Sciences Mathématiques, Université Paris-Sud, Janvier 1987.

[64] H. Nauta. *Ergodicity conditions for a class of two-dimensional queueing problems*. PhD thesis, University of Utrecht, 1989.

[65] A.I. Ovseevich. The Discrete Laplace Operator in an Orthant (algebraic geometry point of view). *Markov Processes and Related Fields*, 1(1):79–90, 1995.

[66] R.N Pederson. Laplace's method for two parameters. *Pacif. J. Math.*, 15(2):585–596, 1965.

[67] I. Plemelj. Ein Ergänzungssatz zur Cauchyschen Integraldarstellung analytisher Funktionen, Randwerte betreffend. *Monatshefte für Math. u. Phys.*, 19:205–210, 1908.

[68] H. Poincaré. *Leçons de Mécanique Céleste*, volume 3. Gauthier-Villars, Paris, 1910.

[69] D. Revuz. *Markov Chains*. North Holland, 1975.

[70] I.B Simonenko. Operators of convolution type in cones. *Math. USSR-Sbornik*, 74:298–313, 1967. English translation in Math. USSR-Sbornik 3, 279-293.

[71] V. Smirnov. *Cours de Mathématiques Supérieures*, volume 3. Éditions Mir Moscou, 1972. Traduit du Russe.

[72] G. Springer. *Introduction to Riemann Surfaces*. Chelsea, New York, 2nd. edition, 1981.

[73] E.I. Zolotarev. *Theory of integral complex numbers with applications to integral calculus*. Doctoral thesis, Moscow, 1874.

Index

Errata to: Random Walks in the Quarter-Plane

Guy Fayolle, Roudolf Iasnogorodski, Vadim Malyshev

Errata to:
G. Fayolle et al., *Random Walks in the Quarter-Plane*,
Stochastic Modelling and Applied Probability 40,
DOI 10.1007/978-3-642-60001-2

Correcting a book is a heavy-tailed process, which has some non Markovian features...

P. 8, line 3 of Definition 2.1.1. Read "$\mathcal{A}_X$" instead of "A_X".

P. 9, line 3 of Definition 2.1.4. Replace "holomorphic mapping" by "continuous mapping (see [34]);".

P. 10, line 3. Delete the extra space after "representing". At the end of formula (2.1.2), replace the dot by a comma.

P. 12. At the end of Proposition 2.1.12, replace the "." (dot) by "X' and Y' being as in definition 2.1.2."

P. 15, two lines after equation (2.2.1). Change "(1.3.6" to "(1.3.6)".

P. 21. In (i) of Lemma 2.3.4: line 1, should be "$|Y_1(x)| \geq 1, \forall |x| = 1$," instead of "$|Y_1(x)| \geq 1. \forall |x| = 1$"; line 3, replaces "is a real analytic curve" by "describes a real analytic curve".

P. 22, line 5. Change "*(i)*" to "(ii)".

- In the formula line -5, replace "$-p_{10}$" by "$-p_{01}$".

P. 24, line following *(iii)*. Read "been" [instead of "seen"].

P. 29, line 6. Change "2.2.3" to "2.2.4".

P. 37, line -8. Replace "automorphism" by "holomorphic mapping".

P. 45. Lemma 3.3.2 is valid only if $d_4 = p_{10}^2 - 4p_{11}p_{1,-1} \geq 0$. Otherwise change $D(x)$ to $-D(x)$ in the displayed formulas (3.3.6).

The online version of this book can be found under
DOI 10.1007/978-3-642-60001-2

G. Fayolle et al., *Random Walks in the Quarter-Plane*, Stochastic Modelling
and Applied Probability 40, DOI 10.1007/978-3-642-60001-2_8
© Springer-Verlag Berlin Heidelberg 1999

- Line -8. Replace "the zeros of y" by "the zeros of g(.)"; line -7, missing colon after "relationship".

P. 46. In the LHS of formula (3.3.8), replace "ω" by "w". Similarly, in the line following this formula, change "where ω" to "where w".

P. 47. The formula defining ω_1 holds only if $p_{10}^2 - 4p_{11}p_{1,-1} \geq 0$. Otherwise, $D(x)$ becomes $-D(x)$.

P. 51, line -1. Replace "$P_0(x,y), P_0(x,y)$ by "$P_0(x,y), P_1(x,y)$".

P. 52, line 3. Change "When π" to "When ρ".

P. 57. In the displayed system (4.2.6). Insert "," (i.e., comma) at the end of the first line; in the third line, read the second "ψ" as "$\widetilde{\psi}$".

P. 62, line -9. Replace "with $h = \alpha$" by "with $h = \beta$".

- Line 13. Read "that" instead of "that that".

P. 66. Line 7. Change "instantiating $\gamma = 1$ in lemma 4.2.4" to "instantiating $\varepsilon = 1$ in lemma 4.2.8"; line 12, "$\mathbb{C}\,\omega$" becomes "$\mathbb{C}_\omega$"; line -10, "using some by" should be "using some of".

P. 72. In the displayed formula (4.3.8). Change "$\prod_{j=0}^{i=1} f_{\delta^i}$" to "$\prod_{j=0}^{i-1} f_{\delta^j}$".

P. 79, line 14. "presentedbelow" becomes "presented below".

P. 89. In the equation line -4. Replace "$\eta(\infty, 0)$" by "$(\eta(\infty), 0)$".

P. 94. Second line of the displayed system (5.1.2), "x^{i-1}" should be "x^{i-L}".

P. 100, line -6. Insert a comma right after "[61]".

P. 110, last equation. Change "$X_1[\overleftarrow{y_3y_4}]$" to "$\overline{X}_1[\overleftarrow{y_3y_4}]$".

P. 111. In the determinant, replace the central term "$p_{0,-1}$" by "$p_{00} - 1$".

P. 112. In the last formulas involving α_2 and β_2: replace "$P_{0,0}$" and "$P_{-1,0}$" respectively by "$p_{0,0}$" and "$p_{-1,0}$". (I.e., P becomes p).

P. 113. In the formula defining T, the second term should be replaced by $+2u(p_{-1,1}p_{0,-1} - p_{-1,-1}p_{01}) + p_{-1,0}p_{0,-1} + (1 - p_{00})p_{-1,-1}$.

P. 114. Replace $Y(x_4) = \dfrac{-b(x_4)}{a(x_4)}$ and $\Re(Y(\infty)) = \dfrac{-b(\infty)}{2a(\infty)}$ by, respectively, $Y(x_4) = \dfrac{-b(x_4)}{2a(x_4)}$ and $\Re(Y(\infty)) = \dfrac{-b(\infty)}{2a(\infty)}$.

P. 116. In line 4 of paragraph **(a)**, "ad" becomes "and".

P. 117, line 2. Change "section 4.7" to "section 4.6".

P. 119. In the line 1 of *Proof*, replace "theorem 5.2.3" by "theorem 5.3.3".

P. 122. Two lines after equation (5.4.12), put "$(X_0(u_k), u_k) \in \mathcal{D}$" instead of "$u_k \in \mathcal{D}$".

- The RHS of formula (5.4.12) should be

$$\psi_1(x) = \sum_k a_k \frac{\pi(x) - \pi(X_0(u_k))}{x - X_0(u_k)},$$

- Last line. Read "since by (5.1.6)" instead of "since".

P. 123. There are several jams in the definitions of the functions. . . !

- Formula (5.4.16) should be

$$\psi(x) = \psi_1(x) + \sum_\ell b_l \frac{\pi(x) - \pi(X_0 \circ Y_0(v_\ell))}{x - X_0 \circ Y_0(v_\ell)}$$
$$= \sum_k a_k \frac{\pi(x) - \pi(X_0(u_k))}{x - X_0(u_k)} + \sum_l b_l \frac{\pi(x) - \pi(X_0 \circ Y_0(v_\ell))}{x - X_0 \circ Y_0(v_\ell)},$$

- In system (5.4.18), replace the expression giving $T(x)$ by

$$\frac{R(x)}{P(x)} \left[\sum_\ell \frac{b_l g(v_\ell)}{(x - X_0 \circ Y_0(v_\ell)) A(X_0 \circ Y_0(v_\ell))} - \sum_k \frac{a_k \pi_0(X_0(u_k), u_k)}{(x - X_0(u_k)) q(X_0(u_k), u_k)} \right],$$

- Right after displayed system (5.4.18), add the following lines:

"and $P(x)$ is a polynomial depending linearly of the coefficients a_k and b_l, which are chosen real, to ensure that $P(x)$ does not vanish in the domain $G_\mathcal{M} \bigcup \mathcal{D}$. This can clearly be achieved in many ways."

- In system (5.4.20), change the first line of definitions to

$$\rho(t) = S(t)\psi(t), \quad K(t) = \frac{G(t)R(t)}{P(t)},$$

P. 124. The factor before the integral in formula (5.4.21) becomes

$$\frac{R(x)H(x)}{2i\pi P(x)S(x)}$$

P. 127. In the formulas giving e_1 and e_2, "ω_2" becomes "ω_1".

- Line -12. Replace "$\frac{dx}{d\omega}$" by "$\frac{dw}{dx}$".

- In the last formula of section 5.5.2.2, change "$(-e_3)$" to "$(w - e_3)$".

P. 130. In the figure representing the real axis, switch the points 1 and x_3.

- Line -14. Change "$i = 1, 2$" to "$i = 0, 1$".

P. 135. In section 6.4.2. Replace (twice) "$X_0 \circ Y_0)^{(n)}$" by "$(X_0 \circ Y_0)^{(n)}$".

- In the formula (6.4.4), change "y(u)" to "y(s)", and "$\eta(s)$" to "$\gamma(s)$" (twice) and add a comma at the end of the formula.

P. 136, line 2. Read "is given" instead of "in given".

- Line -1. Change "(6.4.2)" to "(6.4.4)".

P. 136. In the third fomula from above, replace the "." (dot) by "," (comma), and right after this formula add the following text:

"and $\gamma(.)$ is a fractional linear transform satisfying $\gamma(\eta(s)) = \frac{1}{\gamma(s)}$."

P. 142, line -7. Read "figure 6.5.1" instead of "figure 6.5.2".

P. 143. In formula (6.5.11), change "$y(u)$" to "$y(z)$", and "$\eta(z)$" to "$\varepsilon(z)$".

- Line -3. Replace "$w \to w + \dfrac{1}{w}$" by "$w \to \frac{1}{2}\left(w + \dfrac{1}{w}\right)$".

P. 150, line 7. Replace "5.4 6.5" by "5.4 and 6.5".